Peregrine Falcon

Peregrine Falcon

PATRICK STIRLING-AIRD

FIREFLY BOOKS

Published by Firefly Books Ltd. 2012

First printing

Publisher Cataloging-in-Publication Data (U.S.)

Stirling-Aird, Patrick.
Peregrine falcon / Patrick Stirling-Aird.
[128] p. : col. photos. ; cm.
Includes bibliographical references and index.
Summary: The life of the peregrine falcon including the high places it breeds, feed and roost, its behavior, and how people are protecting this species.
ISBN-13: 978-1-55407-927-8
1. Peregrine falcon. I. Title.
598.96 dc22 QL696.F34S757 2012

Library and Archives Canada Cataloguing in Publication

Stirling-Aird, Patrick, 1943–
Peregrine falcon / Patrick Stirling-Aird.
Includes bibliographical references and index.
ISBN 978-1-55407-927-8
1. Peregrine falcon. I. Title.
QL696.F34S85 2012 598.9'6 C2011-906221-6

Published in the United States by
Firefly Books (U.S.) Inc.
P.O. Box 1338, Ellicott Station
Buffalo, New York 14205

Published in Canada by
Firefly Books Ltd.
66 Leek Crescent
Richmond Hill, Ontario L4B 1H1

Printed in China

Developed by:
New Holland Publishers
London • Cape Town • Sydney • Auckland
www.newhollandpublishers.com

Publisher: Simon Papps
Editor: Marianne Taylor
Design: Roger Hammond at Blue Gum Designers
Cover design: Jason Hopper
Production: Melanie Dowland

Reproduction by PDQ Digital media Solutions Ltd, UK

THIS BOOK IS DEDICATED TO MY WIFE SUSAN,
WHO ENCOURAGED ME TO WRITE IT AND WHO
OVER MANY YEARS HAS SUPPORTED MY VERY
TIME-CONSUMING PEREGRINATING

Acknowledgements

There is an extensive international literature on Peregrines (especially British, other European and American), on which I have drawn in writing this book. As to personal communications, I would like to thank Bob Elliot and Duncan McNiven for information on wildlife crime as affecting the Peregrine, Mike McGrady for pointing me in the right direction in sorting out the subspecies and for general comment on Peregrines internationally, Dave Dick for support in various ways, Gordon Riddle for insight into Common Kestrel breeding behaviour, George Smith and Oscar Murillo for information on a Peregrine PIT tagging scheme, Carol Rawlings and John Turner for data on London and Shropshire Peregrines respectively, and Simon Papps of New Holland Publishers for helpful advice.

The Nature Conservancy Council and subsequently Scottish Natural Heritage have been diligent over the necessary licensing arrangements for detailed survey work. In many cases the owners and occupiers of land visited by me have helped in one way or another, and have shown a welcome interest in Peregrine fieldwork and its results. Going back some four decades, I am grateful for the encouragement shown by John Mitchell, the late Don MacCaskill and Bob McMillan in getting me started on formal Peregrine monitoring. Since then, it has been a pleasure to have worked with many other people in this field (not least Pete Ellis and Hugh Insley, intrepid climbers to Peregrine eyries) and in recent years especially with member organisations of the Scottish Raptor Monitoring Scheme.

Contents

Preface

I first saw Peregrines more than 40 years ago, an adult pair rising and falling in swift flight as they pursued their way along a high, broad-backed ridge in the Scottish Highlands, on one of those summer days of brilliant visibility with the distant summits standing out sharply against the blue skyline. The following February I watched another pair, anticipating the coming of spring, carrying out spectacular courtship flights over the future nesting cliff and dashing to and fro in characteristic powerful flight.

Since then, I have had the privilege of watching many Peregrines and of seeing their headlong flights after prey. However, it was those two early events, and recognition of the bird's importance as an 'ecological barometer' at the top of the food chain, that inspired my involvement in formal breeding season monitoring of Peregrines along the southern fringe of the Scottish Highlands. I was fortunate in coming to such monitoring at a time when the local Peregrine population was recovering from its earlier pesticide-induced decline. Thus there was the excitement of participation in what to some extent was pioneering work, with suspected Peregrine pairs and breeding localities being 'put on the map' in the biological recording sense.

This book is intended to reveal something of the Peregrine's fascinating position in the natural world (with a basic outline of its evolution to the supreme predator that it is) and to describe the main problems that it has faced, especially over the last 200 years. At a time when there is good cause for concern about the degradation of the natural world and the depletion of its biodiversity, I have tried to end on a positive note, the Peregrine having proved hitherto to be a more adaptable, resilient and successful bird than anyone would have forecast just half a century ago. The Peregrine is now one of the planet's best-studied species, but much still remains to be discovered about its way of life, by both scientist and non-scientist, and perhaps by readers of this book.

Opposite: The Peregrine, ever watchful on its lookout post, is a supreme predator in habitats ranging from upland wildernesses to city centres.

1 | The falcon family

The Peregrine (*Falco peregrinus*), soaring on the wind against a cloud-flecked blue sky, stooping down to its home crag from on high or streaking across the void in pursuit of prey or a rival Peregrine, epitomises both the strength and the fragility of nature. The strength comes from the bird having passed through decades of misfortune of one sort or another, emerging with renewed vigour. The Peregrine's fragility is an inevitable consequence of its position as a predator at the apex of a food pyramid, closely linked to the ups and downs of the web of life beneath it in that pyramid. Thus it has been famously described, in a report by Derek Ratcliffe to the British Trust for Ornithology on Peregrines in 1971, as "an ecological barometer of importance to Everyman."

We start with the dry but necessary subject of scientific classification, with the Peregrine placed as a raptor (bird of prey) within the order Falconiformes, one of 29 orders that make up the class Aves (birds). The word 'raptor' is taken from the Latin *rapere*, 'to seize and carry away.' Within the order Falconiformes, the Peregrine belongs to the family Falconidae, and of the 11 currently recognised genera within that family, it is classed in the genus *Falco*.

Going one step back up the classification ladder, the family Falconidae is split into two subfamilies: Falconinae (the 'true' falcons and the falconets) and Polyborinae (the caracaras and the forest falcons). The term 'falcon' derives from the Latin word *falx*, a sickle, said to relate to the shape of the bird's hooked talons (claws). One might suppose that it could equally describe a falcon's flight silhouette, of which more later.

Breaking down this classification further, there are considered to be one genus each of true falcons (genus *Falco*), forest falcons, laughing falcons and pygmy falcons; two genera of falconets; and five genera of caracaras. It's a complex classification, made more so because the geneticists keep changing their minds. Members of Falconiformes differ from birds belonging to Accipitriformes, which include the other diurnal raptors, in various ways. These comprise parts of the skeleton, the bill structure and the sequence of moult of the primary feathers.

Falconiformes are considered to be more closely related to the order

Previous page: the carrion-feeding Southern Caracara (*Caracara plancus*) of South America is a distant relative of the Peregrine, within the family Falconidae.

Opposite: South America is a hot-spot of diversity within the falcon family. It is home to all the world's seven species of forest falcon, including the Lined Forest Falcon (*Micrastur gilvicollis*) which is widespread across the north of the continent.

The kestrels are smaller members of the falcon family which often hover while in search of mammal prey. This is a female Common Kestrel (*Falco tinnunculus*).

Strigiformes (owls) than to other diurnal raptors, with which Falconiformes are thought to have evolved in parallel. The similarities between members of the two orders of raptors (Falconiformes and Acciptriformes) are assumed to have been the result of convergent evolution, whereby creatures with distinctly separate origins pursue similar ways of life, developing broadly similar structures and appearances on the way.

The genus *Falco* has been taken to comprise (apart from the Peregrine) anything from 24 to 39 individual species, depending on how one wishes to divide them. A convenient and at present generally accepted classification gives 38 falcon species in all, ranging in weight from some kestrels at less than 100g (3.5oz) to the heaviest Gyr Falcons (*Falco rusticolus*) at more than 2kg (4.3lb). One or more species of the genus *Falco* breeds on all continents except Antarctica and on many oceanic islands, from 82° N in Greenland to 55° S in Tierra del Fuego at the southern tip of South America. Some falcons have breeding ranges that are geographically separated from areas which they occupy outside the breeding season, while others (in subtropical and tropical latitudes) are resident throughout the year. It is mainly the species breeding in northern latitudes that make annual movements to and from their breeding grounds. Uniquely among the falcons, the Peregrine has a world-wide distribution, covering almost the same ground as the entire *Falco* genus.

Falcons are small to medium-sized raptors. They have long, pointed wings with moderately or very stiff feathers and, in the case of the bird-hunting species, large flight muscles which probably constitute some 12–20 per cent (perhaps more) of total body weight, relatively high wing-loading, i.e. ratio of weight to wing surface area, and short tails. These are features related to fast flight. These birds' way of life depends to a greater extent than that of any other raptor group on wide-ranging aerial pursuit of prey. Therefore, within the raptor tribe falcons merit the description of 'attackers', as opposed to 'searchers' such as the harriers and kites.

Falcon evolution

Where and when did what you might call the ancestral falcons first appear? It has been suggested that South America may have been important in the early story of falcon evolution, due to the present variety of caracaras, forest falcons and true falcons there. A difficulty with this idea is the scarcity of fossil records. Given the available evidence, these are the current lines of thinking on the sequence of events in falcon evolution.

It seems that species with characteristics of the genus *Falco* first existed some 20 million years ago, diurnal raptors as a group having appeared about 20 million years earlier in time than that. It is suggested that forest-adapted species (or possibly scavenging forms), similar to modern forest falcons or caracaras, came first, with the true falcons (which are primarily suited to non-forest habitats) appearing later. The true falcons may have evolved in Africa (more falcon species occur there today than on any other continent) when its open grasslands appeared. The present-day kestrels are thought to be closest in form and habits to the ancestral falcons, with the Peregrine and other 'large falcons' being more recently derived types whose lineage separated from that of other falcons 5–8 million years ago. The current true falcons (with eight or nine separate groups or subgenera within the genus *Falco*, the Peregrine being the only representative of one of these subgenera) may therefore have developed from ancestral kestrel-type generalist predators. The Peregrine itself probably diverged from the other large falcons some 2–2.5 million years before today.

Be that as it may, an early 20th century writer, J. H. Crawford, gave his description (and interpretation) of falcon evolution from the original 'reptile' birds thus: "When the bird first sprung from the reptile, the air was empty of danger. A lumbering flight was good enough, and would probably have been the highest stage reached. A further stimulus was needed. That came in the form of winged enemies; pursuers in the same element, and of their own kindred. From such are the perfection, and all the marvellous mechanism of flight. To distance the pursuer, a bird had to put on all its speed. It could put on no more than it had received. But some were quicker than the rest. The slow perished by slow-winged hawks. The quicker survived, to raise quicker broods. And, so the limits of speed were increased. That there might be no relaxing of the strain, no resting-place in the course of evolution, on the track of the swifter was the swift-winged falcon."

Falcons tend to use their rather short bills rather than their talons for killing prey, for which purpose they have a biting mechanism called the tomial tooth, from *tomium* which in Greek and Latin combined refers to the cutting edges of a bill. The tomial tooth is thus

The Saker Falcon (*Falco cherrug*) is one of the larger members of its family. It inhabits areas of steppe in Eurasia, where its prey comprises mammals and birds.

a serration of the bill, not a tooth as such, comprising a cutting edge on the upper mandible, with a corresponding notch on the lower mandible. Also, falcons' powerful jaw muscles enable them to bite hard into the necks of their prey so as to dislocate the vertebrae. Falcons' feet, with their typical raptorial structure of three forward toes and one rear toe, have evolved basically for grasping and holding prey, not so much for killing it as in the case of the Acciptriformes.

Evolution has given bird-hunting raptors (including the Peregrine) the ability to capture prey larger than they are, the theory being that the

latter are necessarily inferior to their pursuers in acceleration, level speed, rate of climb and manoeuvrability. An evolutionary thrust in the large falcon species has been towards capturing, killing, eating and digesting big, active prey. It seems that the large feet and bills of bird-hunting falcons have developed accordingly, and the tomial tooth may allow them to kill prey considerably larger than would be possible without it.

As a falcon can consume about 25–30 per cent of its body weight in one meal, it must be an advantage to it in having the remainder of a feast to which it can return for second or third helpings, assuming that in the meantime the prey item has not been consumed or taken away by scavengers. Falcons' capacity for big appetites may have two benefits: the hunter can get maximum energy to repay the effort of chasing and capturing large prey; and a sizeable meal provides insurance against starvation at times when subsequent hunting conditions are poor, perhaps over a period of several days.

Reversed sexual size dimorphism is more pronounced in large bird-feeding specialists such as these Gyr Falcons than in predominately mammal-feeding members of the family such as the kestrels. The smaller male is on the left and female on the right.

RSD

Falcons exhibit reversed sexual size dimorphism (RSD), in other words, females are larger than males, in contrast to most bird species where males are larger. RSD is particularly pronounced both in 'true' hawks such as the Eurasian Sparrowhawk (*Accipiter nisus*), and in those falcons like the Peregrine which are specialised bird hunters. A suggested explanation is that larger size is an advantage to the female in her nest-guarding role, while the smaller male is a more agile food provider for the family.

Another parallel explanation put forward is that RSD reduces intraspecific competition for food, allowing males and females to take different-sized prey. They therefore can exploit a wider prey spectrum than would be the case were both sexes the same size. That could be important in the breeding season when nesting falcons are restricted to an area (of whatever size, depending on the species) around the nest site and cannot range as widely in search of prey as they can at other times of the year.

It has also been suggested, perhaps less plausibly, that this difference in size between the sexes may reflect falcons' powerful armaments (their bills and talons) and allow dominance of the male by the female where otherwise conceivably the male could present a threat to the young or to the female herself. Another line of thinking, however, is that larger female size (as opposed to male) may be an incidental result of the need for different-sized sexes.

Nevertheless, if (as is thought) female falcons compete among themselves for a scarce resource (a competent male in possession of a breeding territory) then the scarcer this resource, the bigger the female should be. Given that falcons prey on agile and fast-moving prey, it is probably more difficult to be a high-quality male than it is for species taking food that is easier to collect. Large females may be best able to deter smaller females from consorting with worthwhile males, and thus better able to stake their claims for entry to (or continued existence in) the breeding population. Whatever its cause, it is noteworthy that RSD in raptors increases with the speed, agility and size (relative to the raptor) of their prey.

Opposite: The Gyr Falcon is the world's largest and most powerful true falcon. It has a circumpolar distribution in the Arctic and preys mainly on large birds.

Other northern falcons

Before turning in detail to the Peregrine and in order to show the variety among falcons, it is worth comparing it with two Northern Hemisphere species, namely Gyr Falcon and Merlin (*Falco columbarius*). All three have large circumpolar ranges and are largely or almost exclusively bird eaters but, once again bringing in the esoteric niceties of classification, each can be counted as belonging to a separate subgenus of *Falco*.

The Gyr Falcon (once given also the romantic names Iceland Falcon or Greenland Falcon in its paler colour morphs but Norwegian Falcon in its darker form) is the largest of the falcons. Typical Gyr Falcon weights are 1.05kg (2.3lb) for a male and 1.75kg (3.8lb) for a female. This falcon, a high latitude, Arctic/subarctic bird, preys substantially on Rock Ptarmigan (*Lagopus mutus*) but is a partial feeder on mammals up to the size of Arctic Hares (*Lepus arcticus*). Therefore, it is equipped with relatively shorter and heavier feet than the almost exclusively bird-feeding Peregrine.

The Gyr Falcon, although less manoeuvrable than the smaller Peregrine, is swifter on the wing, has a more sustained flight in pursuit of prey, and is more powerful when striking it. Gyr Falcons generally spend the winter as far north as prey numbers allow, although at times they migrate over considerable distances (for example, to the British Isles) in response to movement of and variation in numbers of their prey, especially Rock Ptarmigan.

As mentioned, the Peregrine too is counted as a large falcon but, by contrast with the Gyr Falcon, it has notably long toes and feet as befits a largely bird hunting existence. Some Peregrine populations are migratory (see chapter 3), but the species in much of its range is more of a stay at home resident than is the Gyr Falcon.

The Merlin, the baby of the trio at some 160g (5.5oz) in weight for a male and 220g (7.6oz) for a female, has been described as a miniature Peregrine and in flight can appear so although with a proportionately slightly longer tail. As a predator on small birds such as the Meadow Pipit (*Anthus pratensis*), the Merlin shows the same aerial predator's compact shape as the Peregrine. A hunting Merlin, however, is more of a contour-hugger than is the Peregrine, typically sprinting low over the moorlands in pursuit of its small passerine prey.

The Merlin has long been seen as having great hunting ability in spite of its small size, hence its use for the once popular sport of lark-hawking, whereby the Merlin pursued the slower but more lightly wing-loaded Skylark (*Alauda arvensis*) in a rising flight. The Merlin enjoyed the medieval name 'Milady's Falcon', denoting the falconry social hierarchy of the time. The Merlin is a mainly migratory species, although some of its populations (as in Britain) remain more or less resident in their general breeding areas, even if outside the breeding season they forsake the hill country for lower ground or the coasts.

Opposite: A diminutive male Merlin, on the alert for small passerine prey, perches on a lookout rock on heather moorland. This is a favourite breeding season habitat of the species.

2 Introducing the Peregrine Falcon

The Peregrine takes the prize for the most extensive distribution of any bird species world-wide, its range extending from Alaska to Patagonia and from north-east Siberia to Tasmania. The species is rivalled in its natural distribution only by the Osprey (*Pandion haliaetus*) and the Northern Raven (*Corvus corax*). The Peregrine, either as a breeder or as a winter visitor, is absent only from desert parts of Africa and Australia, parts of Central Asia and China, part of South America, much of the high Arctic where the Gyr Falcon reigns (although recent less extreme weather conditions there have allowed Peregrines to encroach northwards into the Gyr Falcon's domain) and Antarctica.

The name 'Peregrine' comes from the Latin *peregrinus*, meaning 'foreigner' or 'traveller', describing the bird's wandering, migratory habits in (especially) the northern part of its range. Thus one finds, for example, Wanderfalke in German and Pilgrimsfalk in Swedish. This infers movement over wide areas, and reflects the Peregrine's status as a partial migrant. The scientific name *Falco peregrinus* aptly translates to 'wandering falcon.' As befits its extensive distribution, the Peregrine has been given many local names, such as Duck Hawk, Great-footed Hawk or Rock Peregrine in temperate North America, Tundra Falcon in arctic North America and Greenland, Blue Hawk, Game Hawk or Hunting Hawk (sometimes Blue Hunting Hawk) in Britain, and Black Shaheen ('royal bird') in India.

In falconry parlance the female Peregrine is the 'falcon' (elsewhere in this book I use the term 'falcon' generically) while the male is the 'tiercel'; a tierce, in Latin *tertius*, is a one-third measure. He is described as being about one third smaller than she, although that third is more in terms of weight than of dimensions. The Peregrine has long been a favourite of the falconer, as described in chapter 7.

Previous page: a wandering adult Peregrine ekes out an existence in harsh winter conditions.

Opposite: An adult Peregrine on guard at its nesting crag. Adults are easily visible if facing outwards when against a dark background, less so when perched facing in to the crag with the darker upperparts directed towards the observer.

Flight and speed

The Peregrine, one of the fastest and most aerial of predators, is famous as a supreme hunter of other flying birds. For this purpose it is

endowed with a compact, thick-set body, with long, quite broad but pointed wings and a relatively short tail (both are characteristics of open-country raptors) and very long toes for grasping prey in flight. Along with other raptors, the Peregrine's front inner and rear toes are the two most powerful of the four toes and are armed with the longest talons. The three forward toes and single rear toe work in opposition for a tight grip on prey. The tarsus (lower leg) is relatively short, and is thus well adapted to withstand the shock of high-speed strikes on avian prey species.

Opposite: A Peregrine about to fly from its nest crag. The birds can come and go remarkably quickly and a flight in to or out from the crag may be easily missed.

This sturdily built falcon has powerful chest muscles for swift flight, and is celebrated for its stoop on prey from a height. It is endowed with small bony tubercules in the nostrils which may have some function related to high speed flight, probably helping to maintain correct airflow. The Peregrine is a heavy bird for its size and a consequent feature is a high wing-loading. Nevertheless, it is adept at soaring flight when weather conditions (for instance, sunshine and a good updraft) allow.

Seeing a stooping Peregrine with wings folded back is one of the greatest sights in the natural world.

The Peregrine in flight has a distinctive appearance – 'jizz' in birdwatcher's language – even when seen from some distance. The description "that cloud-biting anchor shape, that crossbow flinging through the air," was aptly given by one writer, John Baker, who spent ten years diligently observing wintering Peregrines in Essex, England. There has been much debate as to the maximum speed at which a Peregrine can fly. Its best speed in horizontal, flapping flight is probably not more than around 105–110km (65–68 miles) per hour. Stooping velocity is another matter and has been put at various speeds. A precise radar measurement produced a speed of 182km (113 miles) per hour. A higher (apparently accurately measured) airspeed in the stoop has shown 300km (186.5 miles) per hour. It has been suggested that the Peregrine can stoop at more than 375km (233 miles) per hour, although mathematical calculations of maximum terminal velocity indicate that this speed cannot be exceeded. Whatever the truth of this (and there are a number of proponents for maximum speeds of this order) it may not be often that a Peregrine gets the chance or has the need to attain such a high speed.

Opposite: The distinctive feature of the immature Peregrine is its brown plumage, and in particular the vertical brown streaks on the underparts, which contrast with the horizontal barring of the adult.

Plumage

The adult Peregrine has striking coloration, blue-grey above and white below (although sometimes buff coloured or even salmon-pink, especially in the female), barred with greyish-black. This dark barring is generally more pronounced in female than in male Peregrines. The adult wears a dark slate colour on the upper part of the head, that colour extending as a still darker 'moustache' from the base of the bill to below the eye. The cere (a bare patch around the bill) and the feet are yellow, although the immature Peregrine's cere is more of a blueish colour. The bill itself and the talons are black. Usually the male sports a whiter breast than the female, but I have seen at least one pair where the reverse was the case.

Above: This Peregrine shows the diagnostic adult features of bluish upperparts and yellow cere, although odd brown immature feathers are retained on the wings.

Juvenile Peregrines are browner (although in some light conditions their upperparts can appear greyish and thus deceptively similar to those of the adults) with vertical streaks of brown, not horizontal barring, on the front. The juvenile's facial 'moustache' is less distinct than that of the adult and its brown upperparts provide good camouflage for well-grown young when they are crouched on the nest ledge. The juvenile, once fledged, shows a conspicuous pale terminal band to the tail, which helps to distinguish it from the adult in flight.

The blue-grey upperparts of adult Peregrines presumably confer some advantage, probably a predatory one. Blue is a rare plumage

colour among raptors, and features only in the bird-eaters. Moreover, the Peregrine's counter-shading, of dark upperparts and pale underparts, may help to disrupt its profile as seen by potential prey species, and thus aid capture of fast and agile prey. From below, the Peregrine's pale underparts are not as obvious against the sky as more sombre ones would be, while its dark rather than pale upperparts are less visible against the ground when viewed from above.

This female in an urban environment, ringed as part of a species study, shows clearly the Peregrine's thickset body and a hint of the large muscles which power the long wings

Size

Female Peregrines are on average 15 per cent bigger than males, and have relatively larger and more powerful feet and talons. Unsurprisingly, size varies somewhat between subspecies and indeed between individual Peregrines of the same subspecies. The larger forms are generally found at higher latitudes, reflecting Bergmann's Rule. This states that northern examples of the same species tend to be bigger than southern ones as larger individuals are better able to stay warm, due to their lesser ratio of surface area (through which heat is lost) to body mass (in which heat is generated).

Typical measurements of these larger forms of Peregrine are, in height, 47cm (19in) for a female and 42.5cm (17in) for a male. Thus the female Peregrine is around the size of a Carrion Crow (*Corvus corone*) although she is about twice its weight. Average wingspans of the larger forms of Peregrine are 97cm (39in) for a female and 84cm (34in) for a male. A Peregrine pair seen in flight together usually gives the impression of a bigger size difference between the sexes than these measurements indicate, and the female has a slightly heavier flight than the male. There is a greater discrepancy in weights among northern birds, average females and average males tipping the scales at 1.125kg (2.4lb) and 680g (1.5lb) respectively, although one Scottish female was found to weigh 1.45kg (3.1lb). At the lower end of the size range, average weights of the Indian subspecies of the Peregrine are some 675g (1.5lb) for a female and 500g (1.1lb) for a male.

Sight

Like other raptors, the Peregrine's eyesight is legendary. Its large, forward-set eyes (dark, as in all falcons) take up a full half of its skull, and give binocular vision for locating prey and accurate judgement of distances to it. The Peregrine has two focal centres (foveae) at the back of each eye, one deep and one shallow. The deep fovea acts like a telephoto lens, and is used to scrutinise distant objects, as the Peregrine turns its head side on. The two shallow foveae provide a binocular perspective, akin to ours. Hence the Peregrine can track three moving objects simultaneously, perceiving a central binocular image and two magnified images, one with each eye.

The Peregrine has a semi-transparent third eyelid, the nictitating membrane, which can be closed to give wind protection at speed. It has been calculated (per Baker again) that, were human eyes to be in the same proportion to the body as the Peregrine's, humans would have eyes 7.5cm (3in) across, weighing 1.85kg (4lb) each. Interestingly, the eyes of the two sexes are said to be the same size, so that those of the smaller male are relatively larger. One can only speculate, but that eye size difference (relative to body size) might reflect more powerful eyesight, which would perhaps confer an advantage on the male Peregrine, as principal provider for his family (including his mate) during much of the breeding season.

There are recorded instances of Peregrines spotting approaching birds, whether prey species or other rival Peregrines, long before the human observer could discern them even with the aid of binoculars. In one such incident the record is of a female Peregrine with large young, shifting around on her perch on the nest crag, calling and watching the sky for a minute or so. The human observer then scanned that same piece of sky with 12x magnification binoculars, and discerned a distant speck which after about a further minute resolved itself into an intruder Peregrine, passing overhead at great speed.

Opposite: Peregrines are particularly vocal when they are on or around the nest crag.

Moult

In common with other birds, Peregrines shed their old feathers and replace them with new ones by moulting, this process taking around 18–26 weeks to complete. Raptors need to maintain flying efficiency (for hunting purposes) during wing moult, so for them moult of the long flight feathers (primaries) is a lengthy process, with only one or two primary feathers being moulted simultaneously. Territorial adult Peregrines do not usually start to moult until breeding is well advanced or completed. In much of the Northern Hemisphere, immature birds probably moult between April and June, i.e. when they are aged 10–12 months. The adult plumage appears at the first moult, so Peregrines seen in this plumage are of any age from two years upwards.

Voice

The Peregrine is a distinctly vocal bird with a repertoire of different calls, the most familiar one being a harsh '*cack-cack-cack*'. This is the alarm call, used against human intruders, potential predators and (in an aggressive context) other Peregrines. There is a so-called wailing call between the two members of a pair, often used by the female to encourage the male to hunt. A more intense version is the whining call. The creaking call (for contact between Peregrines, usually mates) has been variously described. To me it sounds like a slightly musical '*klee-klutch*', often repeated in quick succession. A 'chupping' call is associated with feeding of the young, usually by the female. Young Peregrines utter several of the adults' calls, although less strongly and at a higher pitch.

3 | Peregrine populations and distribution

Previous page: The adaptability of the Peregrine is demonstrated by the way that it has moved into urban environments and flourished in them during recent decades.

Now we need to find our way again through the labyrinth of classification, as 19 geographically separated subspecies of Peregrines are currently recognised. Some of these cover wide areas of one or more continents, while at the other end of the scale three, perhaps now only two, subspecies are restricted to small island groups. As mentioned, there are differences (but sometimes insignificant ones) in size and appearance between various subspecies. Peregrines from regions of high latitude and colder climates tend to be larger than those living in areas of low latitude and warmer climates. Here we have Bergmann's Rule again (see chapter 2).

The individual races of Peregrine are, in alphabetical not geographical sequence: *Falco peregrinus anatum* in North America from south of the tundra zone to northern Mexico; *F. p. babylonicus* in Asia from Iran to Mongolia; *F. p. brookei* in the Iberian Peninsula and the Mediterranean through to the Caucasus; *F. p. calidus* in the Eurasian tundra from Lapland to north-east Siberia; *F. p. cassini* in western South America south from Ecuador to Tierra del Fuego; *F. p. ernesti* in south-east Asia from the Malay Peninsula through Indonesia to New Guinea; *F. p. furuitii* in the Bonin and Volcano islands of the north-west Pacific, although this subspecies may be extinct now; *F. p. japonensis* from north-east Siberia south through Japan to Taiwan; *F. p. macropus* in much of Australia and in Tasmania; *F. p. madens* in the Cape Verde Islands; *F. p. minor* in Africa south of the Sahara; *F. p. nesiotes* in the south-west Pacific from New Caledonia east to Fiji; *F. p. pealei* in coastal areas of western North America; *F. p. pelegrinoides* from the Canary Islands east through inland North Africa to Iraq; *F. p. peregrinator* in Pakistan, India and Sri Lanka to south-east China; *F. p. peregrinus* (as the nominate subspecies, in 1771 the first Peregrine subspecies to be described) in Eurasia south of the tundra zone and north of the Pyrenees, Balkans and Himalayas, from the British Isles to the Russian Far East; *F. p. radama* in Madagascar and the Comoro Islands; *F. p. submelanogenys* in south-west Australia; and *F. p. tundrius* in the Arctic tundra zone of North America and Greenland. It is

Peregrines frequently use man-made lookout posts. Here an adult male Peregrine surveys part of its home range.

curious that the Peregrine is not found as a breeding bird in, for example, Iceland, the Faeroes or New Zealand, where nesting places and prey supplies seem to be plentiful.

Changes in classification

The Peregrine has provided much scope over the years for scrutiny by taxonomists, the specialists in classification. Not so long ago 23 separate subspecies of the Peregrine were recognised, but four of these have been excised from the list to bring the current 'official' number of subspecies down to the 19 listed above.

There has been debate as to whether or not what were called (and are still called by some) the Barbary Falcon *F. p. pelegrinoides* of North Africa and the Middle East and the Red-naped Shaheen *F. p. babylonicus* of Asia should be treated together as one separate species, Barbary Falcon *Falco pelegrinoides.* Uncertainty also surrounds the credentials of the Pallid or Kleinschmidt's Falcon (*Falco kreyenborgi*)

of Tierra del Fuego, at the southern tip of South America. Today these are taxonomically 'split' by some authorities, while others still treat all three as Peregrines, of the subspecies *Falco peregrinus pelegrinoides, F. p. babylonicus* and *F. p. cassini* respectively. It has been argued that, in particular, the *babylonicus-pelegrinoides* complex of falcon populations meets the biological criteria for species separation, since these are desert-adapted falcons, and nowhere else has the Peregrine evolved a distinctive desert form.

In all probability, there will be further revisions in Peregrine subspecies classification (this is fertile ground for the taxonomy enthusiast). For example, recent findings suggest that the partitioning of *Falco peregrinus anatum* and *F. p. tundrius* in North America cannot be substantiated. Individuals from these two alleged Peregrine subspecies have been found to be genetically indistinguishable, on a current and a historical basis. The suggestion here is that the subspecies *Falco peregrinus tundrius* should be treated as one with and subsumed into *F. p. anatum.*

Opposite: Dark upperparts and orange underparts make the south Asian *peregrinator* subspecies of the Peregrine highly distinctive. Alternative English names for this form include 'Black Shaheen'.

Migratory differences between subspecies

Peregrines nesting in northern latitudes are highly migratory; those populations breeding furthest north winter furthest south. Thus, Arctic-nesting Peregrines are trans-equatorial migrants in both the New and the Old Worlds. Peregrines breeding in the American Arctic and in Greenland travel south on migration to Central and South America, some as far as Argentina. Breeding birds from the Eurasian Arctic have their winter quarters in Europe, Africa, the Middle East, India, China and south-east Asia.

Peregrines nesting at lower latitudes in the Northern Hemisphere (as in the British Isles) are basically resident at or near their breeding areas throughout the year, although there can be some movements, relatively local and at times weather-inspired. Typically these movements are from mountainous country to low ground or to coastal locations in winter.

There seems to be a tendency for adult male Peregrines especially to remain on their breeding territories. The reason presumably is that, although theoretically the larger female is the better equipped of the pair to cope with winter conditions and thus to stay on the territory, the male cannot afford to leave it for more congenial winter quarters elsewhere lest he lose possession of his territory to a rival male. Wandering males will be ever on the alert for opportunities to stake a claim in the breeding population and thereby to get the chance to pass on their genes to succeeding generations of Peregrines. There is some migration northwards of Peregrines from the breeding populations in the Southern Hemisphere.

Opposite: When Peregrines breed in wilderness habitat – such as here in the Scottish Highlands – they often nest along a steep-sided river valley. Although nest sites may be plentiful, prey species can prove to be relatively scarce.

Peregrine populations

Peregrine numbers world-wide have been assessed with varying degrees of accuracy and commitment. Clearly, it is generally easier to obtain complete or near complete population figures in small, sometimes densely peopled, countries than it is in larger, less inhabited areas, some of which have a claim to be described as true wilderness. As is the case with most other raptors, the counting unit must usually be the territorial, breeding pair. It is these Peregrines that are tied to their nest-site areas in the breeding season, and form a geographically fixed segment of the overall population, for that period of the year at least. Consequently, Peregrine numbers are usually expressed in terms of territorial (or breeding) pairs. It is generally much more difficult, if not impossible, to get a handle on overall numbers, breeders and non-breeders, as the latter are genuinely wanderers – hence again the species name Peregrine. However, good counts of both sectors of the population can be made at concentration points of migrating raptors.

Some examples of recent breeding population estimates over wide geographical areas are as follows: in Europe, 12,000–25,000 pairs; in the United States (including Alaska) 3,000 pairs; in Greenland, 500–1,000 pairs; and in Australia, 3,000–5,000 pairs. For Canada, a calculation (of individual birds, not breeding pairs) showed that in 2005 there were at least 969 mature *Falco peregrinus anatum* and 199 mature *F. p. tundrius* individuals. It was thought that in reality numbers were higher

(especially of the *tundrius* subspecies); a considerably higher population level seems plausible.

These are all very broad estimates and it is important to note that figures for a number of other monitored Peregrine populations are likely to be pitched too low, in some cases markedly so. The reasons for this are less than complete recording and/or recent increases through world-wide colonisation of man-made sites (quarries, bridges and large buildings) by nesting Peregrines, together with recovery in numbers following decreases during the organo-chlorine pesticide era and lessening of deliberate persecution by mankind (see chapters 7 and 8).

There are no continent-wide population estimates for either Africa or Asia, although one can pick out quite precise figures for some of the more thoroughly surveyed national populations in these continents. For instance, Peregrine numbers were put at 400 pairs in South Africa (in 2000) and at 350–400 pairs in Zimbabwe (in 1992) respectively; and there were estimated populations of 70 pairs in the Malay Peninsula in 2007 and 25–30 pairs in Turkmenistan in 2005.

A catalogue of some better known and more thoroughly surveyed Peregrine populations, in this case in Europe and tracking broadly from west to east, brings out these figures: in the Republic of Ireland, 265 pairs; in the United Kingdom of Great Britain and Northern Ireland and the Isle of Man, 1,437 pairs; in Portugal, 50–100 pairs; in Spain, 2,000–2,500 pairs; in France, 1,150–1,250 pairs; in Belgium, 24–26 pairs; in the Netherlands, 50 pairs of which nearly all nest on man-made structures; in Switzerland, 200 pairs; in Denmark, 3 pairs nesting on natural features but 6–8 pairs overall; in Norway, 800–1,000 pairs; in Sweden, 150–175 pairs; in Italy, 1,100–1,400 pairs; in Germany, 1,000 pairs; in Croatia, 160–200 pairs; in Austria, 224–324 pairs; in the Czech Republic, 30 pairs; in Slovakia, 91 pairs; in Hungary, 12 pairs; in Finland, 298 pairs; in Bulgaria, 120–180 pairs; and in European Russia, 1,500–2,000 pairs.

These are merely snapshots of some European Peregrine populations for which figures are available. They encompass nation states in which there have been comprehensive national surveys of the species, and others where, while there has been at least some local or regional Peregrine monitoring, precise numbers are not known for the national populations in question. Moreover, in some instances different sources of national information used in these assessments give varying population figures within individual nation states.

Note also that separate timeframes (but basically spanning the first decade of the 21st century) apply for these estimates of the Peregrine's European status. Comparing these relatively detailed calculations of Peregrine numbers in Europe with the quoted very broad (2004)

Opposite: The Peregrine subspecies *pelegrinoides* of North Africa and the Middle East is also known as the Barbary Falcon. Increasingly authorities are giving it the status of a full species in its own right, and it is sometimes treated as a 'split' together with the Peregrine subspecies *babylonicus* of Asia.

calculation of 12,000–25,000 pairs continent-wide, clearly there are approximately 3,000– 16,000 'missing' pairs. A reasonable assumption is that the total population is a good deal higher than the lower estimate of 12,000 pairs, but how can one determine this without some form of co-ordinated pan-European Peregrine survey, and would the resources for this be forthcoming? Europe is well supplied with keen raptor enthusiasts (a European Peregrine Falcon Working Group was established recently) and with ornithologists generally, by comparison with some other parts of the world, so such a survey is a possibility. However, with a less than complete inventory of Peregrine numbers in Europe at present, it is not

The Peregrine's mountainous domain in Norway where, as elsewhere in Europe, significant population recovery has taken place in recent years.

surprising that the picture is more shadowy for some other continents.

As in much of Europe, the Peregrine has been closely studied in the United States and Canada where its recent history is a well-catalogued one of serious decline (and indeed extinction in a large part of the species' range there), followed by a remarkable recovery. This story is dealt with partly here and partly in chapter 8. A United States census was carried out in 2003, when 438 Peregrine territories were checked. This sample size led to an estimate of 2,435 pairs for the United States overall. A breakdown between the U.S. Fish and Wildlife Service's administrative areas flagged up the following regional populations: Pacific, 472 pairs; Rocky Mountain/Great Plains, 367 pairs; Southwestern, 260 pairs; Midwestern/Northeastern, 315 pairs; Southeastern, 21 pairs; and Interior Alaska, 1,000 pairs.

Recovery of American populations

On the basis of Peregrine territories actually monitored, the North American population as a whole has seen a 2,600 per cent increase over the last 40 years. Much of this success story, as the species recovered from the insidious effects of organo-chlorine pesticides, was due to the efforts of The Peregrine Fund, founded by Professor Tom Cade of Cornell University, Ithaca, New York, in 1970, at a time when actual extinction of the Peregrine in the wild was feared. That project received its initial impetus from the key International Peregrine Conference, held at the University of Wisconsin in 1965, at which the full extent of the problems facing the Peregrine were set out in no uncertain terms.

The Peregrine Fund (now metamorphosed into the World Center for Birds of Prey at Boise, Idaho) carried out a highly successful programme of Peregrine captive breeding (from 1971) and release (from 1974), which broke new ground in the conservation of this and other raptors. The two objectives were to re-establish the extirpated Peregrine population of the eastern United States, and to augment its much reduced western population. These objectives were met; to take an example from just one state, New York, the Peregrine population may be increasing still, with 62 territorial pairs recorded in 2006 and 67 in 2008, an excellent recovery from actual or virtual extinction in the early 1960s. The *anatum* and *tundrius* subspecies of the Peregrine were put on the United States Federal Endangered Species list in 1970, but were taken off that list by 1999.

A two-stage Peregrine recovery programme (of population survey and of captive breeding and reintroduction) also operated in Canada from 1970, concentrating on the threatened *anatum* subspecies in the south and east of that country. Under that programme, captive breeding of Peregrines began in 1972, releases started in 1975 and the first captive-reared birds bred in the wild in 1977. In 1978 Canadian Peregrines of the *anatum* subspecies were given the status of 'Endangered', while the *tundrius* and *pealei* subspecies were classed as 'Threatened' and 'Rare' respectively. In 2007, the Committee for the Status of Endangered Wildlife in Canada downgraded the status of all Peregrines in Canada to 'Special Concern.'

Notwithstanding this much improved picture, there seems to be scope for further recovery of the Peregrine breeding population as a whole in North America and in Greenland, although some areas (such as interior Alaska, perhaps) may be saturated already. The North American/ Greenland Peregrine population was estimated at somewhere between 10,600 and 12,000 pairs before the pesticide-induced decline, all but 2,000–3,000 pairs of which are thought to have occurred in the boreal (northern) and arctic zones above 55° N.

City-centre habitats, such as here in New York City, provide secure, high-rise nest sites for Peregrines and an abundant food supply in the form of Feral Pigeons.

New (and old) habitats

The Peregrine's increasing use of buildings and other man-made structures for nesting has given a welcome boost to the species' breeding populations, in Europe, North America and elsewhere. Cathedrals, office tower blocks, bridges and industrial structures (even very noisy ones) have all been used, showing the adaptability of a bird that was traditionally regarded as a crag nester.

While the Peregrine's use of such artificial 'cliffs' has been a first-rate

success story, re-establishment of the previously extensive tree-nesting population of central and eastern Europe (reckoned to number some 4,000 pairs formerly) has been slow to get off the mark. Only around 20 pairs of tree-nesters have returned (by way of reintroduction in eastern Germany) so most of the former range, covering also the lowlands of Poland, Russia, Belarus and the Baltic States, has still to be re-occupied. Further large-scale reintroduction of tree-nesting Peregrines is needed there.

Looking on the bright side again, in terms of total numbers a current world-wide estimate is of an overall adult population of

1,200,000 Peregrines, in this case not pairs but individual birds. Therefore, in calculating total global numbers of the species, one would have to take into account an additional but unknown number of immature Peregrines.

Opposite: Man-made structures, including industrial ones such as electricity pylons, offer Peregrines useful perching places as well as nest sites.

Population status today

As in other fields, nature conservation deals in categories, conventions, schedules (of species) and the like. This brings us on to another form of classification, the process of listing under criteria established by BirdLife International, formerly the International Council for Bird Preservation. Although the Peregrine still faces distinct threats (see chapters 7 and 8) it is treated by BirdLife International as a species of 'Least Concern', the lowest of five categories of what one might call alert status. BirdLife International's three global criteria that keep the Peregrine out of the next category up ('Near Threatened' above which are the categories 'Vulnerable', 'Endangered' and 'Critically Endangered') are as follows: first, the species has a very large range so does not approach the threshold for the 'Near Threatened' category under the range size criterion; second, the Peregrine's population trend seems to be stable, hence it falls short of the 'Near Threatened' population trend criterion; and finally, its numbers are such that it comes below the threshold for 'Near Threatened' in terms of the population size criterion.

Narrowing the focus down somewhat, and again looking at BirdLife International criteria, the Peregrine's original European listing (based on 1970–1990 trends) was as a 'Species of European Conservation Concern' within the SPEC 3 category. This covers those species having (in technical terms) an unfavourable conservation status, but whose global populations are not concentrated in Europe. At the same time, the Peregrine had the European threat status of 'Rare.' The second European listing (taking into account 1990–2000 trends) removed the Peregrine as a 'Species of European Conservation Concern' and gave it the welcome European threat status of 'Secure.'

The Peregrine is therefore in a happy position among European raptors, more of which declined than increased over the decade 1990–2000. Bringing the focus in closer to the United Kingdom, there is a system of red, amber and green listing under 'The population status of birds in the UK: birds of conservation concern', drawn up by various government agencies and non-government conservation organisations. As the colours indicate, these categories imply species of highest, moderate and least conservation concern. The Peregrine, formerly on the amber list, was given green status on this list in 2009, reflecting its population recovery in the United Kingdom.

4 | Homes and meals

The Peregrine's basic requirements are open country, an adequate food supply (primarily, in the form of avian prey) and sheer or at least steep rock faces, either inland crags – including river gorges – or coastal cliffs, for nesting. To these natural nest sites one must now add, increasingly, man-made structures (among them quarries, both disused and working) if big enough; what are these to a Peregrine other than attractive, secure cliff substitutes?

Disappearance of much of the original natural forest cover (through man's intervention) in various parts of the world led to open country opportunities for the Peregrine, in terms of hunting space, suitable prey species and probably additional nest site availability on crags, especially small ones, that had been screened by forest cover before. One can speculate that in Britain, for example, by the time of the Middle Ages nesting and hunting conditions may have been ideal for the Peregrine. A large proportion, probably most, of the natural woodland had gone by then and the non-intensive agriculture of the time allowed plenty of good habitat for the Peregrine's open-country prey species.

Outside the breeding season Peregrines have scope to range much further away from their nesting areas; such wider-travelling birds are probably mainly those which breed in the far north of both the New and the Old Worlds. Although the Peregrine is primarily a crag nester, some tree-nesting populations are found, as in various European countries (although much reduced there as described in chapter 3) and in parts of Australia. Other birds' disused tree nests make good breeding platforms for Peregrines if they are large enough, for example those of Northern Raven in Europe, Bald Eagle (*Haliaeetus leucocephalus*) in the United States and Wedge-tailed Eagle (*Aquila audax*) in Australia. In northern regions of Eurasia and North America, earth banks of rivers or even the ground are used for nesting. In some parts of the Baltic, for instance, breeding takes place on hummocks among the pool-studded bog country there. In more arid areas the Peregrine chooses river canyons and dry wadis (ravines or valleys which in the rainy season become watercourses) as well as exposed rock formations for nesting.

Previous page: A Peregrine in the Scottish Highlands stands guard over its prey, in this case a Red Grouse

In really dry environments the Peregrine is replaced by other falcon species, the Lanner Falcon (*Falco biarmicus*) in Africa, the Saker Falcon (*Falco cherrug*) in central parts of Asia and the Prairie Falcon (*Falco mexicanus*) in western North America. These falcons are mammal-hunters as well as bird-eaters, and thus have a competitive advantage over the Peregrine in the harsh environments that they inhabit. The Peregrine is rare in tropical areas and is dependent there on gaps in the forest cover.

Suitable coastal breeding habitat is highly sought after by Peregrines thanks to a combination of safe cliff nest sites and plentiful seabird prey.

An immature Peregrine feeds on a Starling (*Sturnus vulgaris*). Note that it has chosen an elevated perch from which to do this so that it can look out for potential danger.

Prey range

The Peregrine merits the description of bird-hunter par excellence. In Britain alone, at least 137 different bird species are known to have been taken as prey by Peregrines, ranging in weight from Great Black-backed Gull (*Larus marinus*) at 1.47kg (3.2lb) to Goldcrest (*Regulus regulus*) at 6g (0.2oz) – the weight of the very small five pence sterling coin. Among these prey species at times have been the Peregrine's raptor kindred, Hen Harrier (*Circus cyaneus*), Common Buzzard (*Buteo buteo*), Eurasian Sparrowhawk, Merlin and Common Kestrel. Even such difficult targets as the Common Swift (*Apus apus*) are caught occasionally.

The list of recorded Peregrine prey species from Central Europe is even longer, totalling 210. Moreover, it has been estimated that world-wide between 1,500 and 2,000 bird species have been taken as prey by Peregrines. The range of prey typically taken is generally much smaller, though, and is normally gathered from the more common birds in the individual falcon's hunting area, for example moorland birds in the hill country and seabirds at coastal haunts.

In Britain and Ireland at least, Domestic or Feral Pigeons (*Columba livia*) feature prominently as Peregrine prey, especially in the breeding season. Indeed, over much of the Peregrine's range various pigeon species are the favourite, perhaps sometimes the most readily available, prey. There is a theory that the soft plumage of pigeons evolved to give them a slight extra edge in escaping from the clutches of pursuing falcons. Some Peregrines (whether individual pairs or, on a much larger scale, separate subspecies) may concentrate their hunting efforts on a few prey species, at least at certain times of the year. For instance, one Scottish pair took a liking to Black-headed Gulls (*Larus ridibundus*) to feed to their young. The North American coastal subspecies *Falco peregrinus pealei* is said to prey largely on four species of seabird in parts of its range. When the breeding season is over Peregrines are particularly attracted to estuaries, for the flocks of waders (shorebirds) and wildfowl wintering there.

A variety of large birds, including Black-throated Diver (*Gavia arctica*), Grey Heron (*Ardea cinerea*), Greylag Goose (*Anser anser*) and

Northern Raven, have been seen to be struck down by Peregrines although not necessarily killed by them in the process. In some, perhaps many, instances the birds were probably attacked through aggression and were not necessarily treated as prey items as such. While Peregrine attacks on Northern Ravens in particular usually seem to be motivated by aggression, there are three recorded instances of juveniles of this species taken as prey to Peregrine nests in the Scottish Highlands.

I once found plucked primary feathers of a Pink-footed Goose (*Anser brachyrhynchus*) or Greylag Goose a few hundred metres out from a Peregrine nest crag, in a location where Golden Eagle (*Aquila chrysaetos*) – the only other likely predator on geese – is not seen, at least not seen on any regular basis. Seton Gordon, a prolific early 20th century writer on natural history and folklore in the Scottish Highlands, described vividly a Peregrine assault on four Greylag Geese; one goose was separated from the other three and was forced to the ground by the Peregrine which then pursued the three geese, struck one down and chased after the remaining two, at which point the proceedings were interrupted by a Golden Eagle which stooped at the geese and then disappeared rapidly, mobbed by the Peregrine in hot pursuit. Neither of the two grounded geese had been killed, although the Peregrine had seized one by the neck and had used the typical falcon bill action on its head.

Food intake

Daily food requirements of wild Peregrines have been estimated, approximately at least, from known food intake of captive birds. For these, typical average daily consumption is 100–110g (3.4–3.8oz) for a male and 140–150g (4.8–5.2oz) for a female, the latter having a slightly lower food requirement in proportion to her bulk than the male due to larger overall body size and thus relatively lower heat loss. There is no reason to think that amounts of food needed by wild Peregrines differ much from these averages. If so, on a daily basis wild male and female Peregrines would have to eat prey equivalent in size to a Mistle Thrush (*Turdus viscivorus*) and a Golden Plover (*Pluvialis apricaria*) respectively.

Contrast the male Peregrine's food requirements against the weight of an adult Red Grouse, the British and Irish subspecies of the Willow Ptarmigan (*Lagopus lagopus*), at around 630g (1.4lb). This is about the heaviest bird likely to be taken by a British or Irish male Peregrine. One can deduce that (allowing for wastage at 20 per cent of total prey weight for any carcass not picked clean) a single Red Grouse would sustain its Peregrine captor for five days. Likewise, a Mallard (*Anas platyrhynchos*) weighing 1.01kg (2.2lb) would feed a female Peregrine for five days. Interestingly, Dugald Macintyre, a falconer-gamekeeper writing in the middle of the last century about Kintyre in the south-west Scottish Highlands, mentioned a Peregrine returning to feed on a Mallard on the fourth day after the kill had taken place. Of course, that could have

A Peregrine of the subspecies *minor* in Ethiopia carries prey back to the nest to feed to its young.

happened only when no sharp-sighted or keen-nosed scavenger had come on the scene to steal the falcon's kill in the interim.

From time to time extravagant claims have been made, often by those who are less than well-intentioned towards Peregrines, as to the numbers of birds 'slaughtered' by them. Macintyre wrote also: "I should say [for] a wild peregrine working a wide country of moor and loch and arable, the bill of fare for a week is something like this: Monday, mallard; Tuesday, second feed from same mallard; Wednesday, one green plover [Northern Lapwing]; Thursday, [Red] grouse; Friday, second feed from same grouse;

Saturday, curlew; Sunday, second feed from same curlew." Here Macintyre was referring to some of the more common prey species in his area, and in this piece seems not to have singled out the Peregrine as a 'culprit' although in his working life he did not allow it to go unmolested. Taking into account his several writings, clearly Macintyre was a keen observer of the wildlife around him, but even for a female Peregrine the week's kill list as described would represent about twice the individual bird's known food needs, allowing for the same 20 per cent wastage for a carcass not picked clean.

Opposite: A stooping Peregrine can reportedly attain speeds of up to 200 miles per hour.

Ratcliffe put into perspective the likely extent of Peregrine predation, in terms of weight of prey killed on an annual basis. He calculated that throughout the year an adult Peregrine pair would take total prey weighing 116kg (256lb), while an average-sized brood of 2.5 young reared by the pair (see chapter 5) would need another 118kg (260lb) of captured prey in that same year, from hatching through to (and including) independence, and assuming of course no post-independence juvenile mortality. Again, these calculations allow for 20 per cent wastage. In order to obtain this total of 234kg (516lb) of captured prey requirement, in most situations a Peregrine pair (and their initially dependent and then independent young) will draw on a quite wide range of prey. Thus predation impact will be spread across what is usually a broad spectrum of prey species.

Studying Peregrine diet

It is in the breeding season that one can best obtain an idea of the Peregrine's choice of prey from within such a spectrum. Then the remains of kills are more likely to be seen (at or near the breeding crag) than at other times of the year when the bird is ranging more widely or, in the case of migratory races of the Peregrine, is entirely absent from the breeding grounds. The plucked feathers from the Peregrine's victims can be quite obvious, especially those of pale-plumaged birds. Notches bitten out of the breast-bone are indicative of predation by Peregrines rather than by other, particularly mammalian, predators. Casual sighting of prey remains, however, gives only a superficial picture of overall prey intake.

Overleaf: an immature Peregrine's stoop causes panic among a flock of Starlings.

Detailed nest inspections are needed for a more comprehensive assessment of the Peregrine's choice of prey; see chapter 9 on the legalities of this.

Outside the breeding season, identification of prey species is inevitably somewhat hit and miss, with much more of the evidence being found far and wide, although favourite Peregrine perches (where the bird plucks and consumes prey) may provide clues. The Peregrine's pellets or castings (regurgitated masses of undigested prey remains, largely feathers), deposited at perching and roosting places, provide further information as to its choice of meal. Identification of feather remains in pellets is an acquired skill, but domestic pigeon rings found packed in pellets leave no room for doubt.

Another item occasionally found in Peregrine pellets is 'rangle', a falconry term for small stones eaten by the birds to help their digestion. A falconer's ancient saying, quoted by E. B. Michell in 1900, recommends: 'Washed meat and stones makyth an hawk to fly.' Although rangle is associated mainly with captive birds, I found a pellet near a Peregrine nest crag that contained several pieces of quartz measuring 13–19mm (0.5–0.75in) across,with ingested feathers attached, and on another occasion a pellet containing smaller stones.

Usual and unusual prey

At times the Peregrine makes some unexpected prey choices. In Scotland and northern England, overland movements of Manx Shearwater (*Puffinus puffinus*), Knot (*Calidris canutus*), Kittiwake (*Rissa tridactyla*) and Common Tern (*Sterna hirundo*) have been intercepted by Peregrines. These unusual (for inland-nesting Peregrines) prey items were apparently caught at or near the falcons' breeding places, and not by long-distance hunting flights to the prey species' coastal habitats. Of more importance is the spring and autumn bonanza of migrating Fieldfares (*Turdus pilaris*) and Redwings (*Turdus iliacus*), on their way to and from their Scandinavian breeding grounds, inviting the attention of Peregrines along the way.

Probably less frequently, the Peregrine (along with some other raptors) turns cannibal. There are at least three records, from both England and Scotland, of Peregrine-on-Peregrine predation, including one of an adult female (with fledged young at that stage) feasting on another adult female. What may have occurred here is territorial aggression ending fatally, rather an initial attempt to treat another Peregrine as prey.

Peregrine predation on mammals is a minority activity for the species. Nevertheless in Britain small Rabbits (*Oryctolagus cuniculus*) and the young of Mountain Hares (*Lepus timidus*) turn up from time to time as prey on Peregrine nests. I was surprised one year to find a half of a freshly killed Mole (*Talpa europaea*) within 20 metres of a Peregrine nest ledge with two recently fledged young nearby. Conceivably, however, the Mole had not been killed by a Peregrine but had been pirated from Common Buzzards breeding in the vicinity, or from Common Kestrels nesting only 50 metres away. Occasional carrion feeding by Peregrines has been noted, generally in winter and in severe weather conditions.

Increasing use of towns and cities by nesting Peregrines has provided

The benefits of breeding by the coast include a plentiful supply of seabird prey for the Peregrine. Here a young Arctic Tern (*Sterna paradisaea*) is the target.

many opportunities to study their feeding preferences in these man-made habitats. One such study took place in Bristol, Bath and Exeter in England, between 1998 and 2007. Before then it had been thought that city-dwelling Peregrines fed mainly on urban species such as Feral Pigeon. Peregrines roosted on tall office buildings in Bristol and nested in Bath and in Exeter from 2006 and 1997 respectively. Each of these cities is surrounded by a wide range of habitats. During this study Peregrine prey remains (from a total of 98 species) were collected at regular intervals from roosting and breeding sites in these cities. Feral Pigeon was the most important prey species, at 42 per cent of prey by numbers although 63 per cent by weight. However, the study found that the diet of these urban Peregrines was very varied, that they regularly hunted prey species that are rarely found in urban environments, and that much nocturnal hunting takes place.

These findings were supported by smaller scale studies in other

English cities and by various research results from continental Europe. Similarly across the Atlantic, observations from the Empire State Building in New York have revealed that Peregrines there regularly hunt nocturnal migrants. It seems that at night urban-hunting Peregrines are helped by artificial light directed at buildings, and that in these circumstances the pale underparts and dark upperparts of some migratory species passing over urban areas show up well from below, although (as nature intended) they provide good camouflage from above.

Hunting techniques

The next step is to look at the way in which the Peregrine catches its prey. As the bird is an open-country raptor observing its hunting techniques is easier than with raptors in closed habitats such as woodland. Some falconry terms are useful when describing the Peregrine's hunting way of life, namely 'ringing up' for rising in wide or narrow circles, 'waiting on'

When the Peregrine is in level flight it is still a force to be reckoned with but the Arctic Tern's chances of escape are slightly better.

for soaring steadily overhead and 'pitch' for the highest point that a falcon reaches when waiting on.

What has been described as still-hunting (not a falconry term) is a frequent technique used by Peregrines, and one that saves energy as the hunter need only fly off its perch, usually high on a cliff, once potential prey is sighted. It is noticeable how often the male Peregrine, as the one that must 'bring home the bacon' for his mate and family, stays on as high a lookout post as possible in order to spot prey at a distance. At the same time he is probably equally on the alert for intruder male Peregrines.

Some studies have shown that Peregrines are less successful hunters when waiting on than they are when still-hunting, on the basis that the intended prey may well spot a falcon on its pitch and thus have a better chance of avoiding its attack. Nevertheless, one might speculate that some potential prey individuals may be familiar with the layout of a Peregrine territory and the locations of their enemy's favourite perches, thus managing to spot a still-hunting falcon before it launches its attack. On the other hand, research involving a Peregrine pair nesting on a power plant in Alberta, Canada, indicated that the male bird of the pair was more successful with waiting on hunting than with still-hunting, the reverse being the case for the female.

The Peregrine will stoop down from its pitch at quarry or, if still-hunting and depending on relative positions of predator and prey, will also stoop on the quarry, chase it in more or less level flight or fly up and perhaps to some distance away (before turning back again) so as to convert height into speed for its attack. If successful in its hunt, the Peregrine will seize the prey in mid-air (sometimes employing the head- or neck-biting action) or will knock it to the ground with clenched or partly open talons. Unsurprisingly, coast-nesting Peregrines seem to take care to avoid dropping birds that they have caught into the sea.

If a Peregrine starts its hunt from a low level it may ring up, driving the target upwards to try to get above it and to prevent it from seeking safety on the ground. However, ground level may not necessarily be a safe place since, unless there is suitable escape cover, the Peregrine may be able to chase its victim to and fro, tire it out and eventually seize it. Clearly Peregrines are successful sometimes in taking prey on the ground, as when the unfledged young of various ground-nesting birds are captured. I have twice seen from close quarters Peregrines on the ground searching among Ling Heather (*Calluna vulgaris*) and on each occasion obviously trying to catch an adult Red Grouse. Another time and from a greater distance I watched a male Peregrine jumping and fluttering for about 15 minutes over a slight hollow, about 10 metres across and filled with Bog Myrtle (*Myrica gale*) and Ling Heather. Close inspection revealed a Common Sandpiper (*Actitis hypoleucos*) apparently hiding in the centre of the hollow.

Various estimates have been made of the Peregrine's hunting success rate, i.e. the proportion of hunts ending in a kill. These estimates vary from a success rate of 8 per cent to one of almost infallible capture of prey. The truth is probably somewhere between these two extremes. Much may depend on the Peregrine's genuine need to kill, from hunger or in order to feed its mate or young. Richard Treleaven spent more hours watching hunting Peregrines than perhaps anyone else has done, in his case at the Cornish sea cliffs in England. Treleaven distinguished between what he termed high-intensity and low-intensity hunting. With the former the Peregrine means it for real and Treleaven found that 69 per cent of such hunts ended in prey captures. Of these successful high-intensity hunts, 62 per cent were still-hunts and 38 per cent were waiting on ones. In the case of low-intensity hunting the Peregrine presumably does not need to kill but cannot resist the temptation to chase prey species (perhaps just for flying practice) but does not mean to secure them as food.

Could there be an inherent tendency for Peregrines (and other raptors) not to 'waste' prey by depleting their populations unnecessarily, since it is on these populations that the raptor's survival ultimately depends? Other observers have noted the same phenomenon of apparent low-intensity hunting by Peregrines. I saw what seemed to be a striking example in the Scottish Highlands, when a female Peregrine came off her nest crag and circled up with the male, both birds then flying off together to converge on a Domestic or Feral Pigeon heading past and away from the crag. All three birds then flew on in formation for several hundred metres, the Peregrines together and at a short distance (I guessed about 20 metres) from the pigeon. First the male Peregrine and then the female made what seemed to be half-hearted feints at the pigeon which dodged and dived down, the female stooping at it once before gliding back to her nest crag.

When watching hunting Peregrines in Cornwall, Treleaven had the advantage of the wider fields of vision that coastal locations provide, as compared with inland ones where the birds all too readily disappear round the side of a hill. He considered that a still-hunting Peregrine spots its potential prey 3.2–4km (2–2.5 miles) away. Treleaven wrote dramatic accounts of still-hunting directed against passing Domestic or Feral Pigeons and described combined interceptions by both male and female Peregrine working together, with one or other of the pair then making the actual kill.

Treleaven felt that a hunting Peregrine relies a great deal on the element of surprise, often stooping at prey from above, passing underneath it and then coming up in a blind spot from beneath the prey. Baker echoed that observation, adding: 'Some soaring peregrines deliberately stoop with the sun behind them. They do it too frequently for it to be merely a matter of chance.'

Evading Peregrine attacks

It might be thought that the odds are firmly loaded against the Peregrine's prey species, but the hunted have some advantages. Many such species are more manoeuvrable (although slower) than the falcon. Staying motionless is one anti-predator technique as, like many other raptors, the Peregrine is particularly sensitive to movement. Flocking of prey species is another such strategy, as waders do on estuaries. The aim there is to confuse the predator with an abundance of closely packed prey. That strategy may have its limitations, as shown by a study of Peregrine predation on Dunlins (*Calidris alpina*) at an estuary in British Columbia, Canada. There Peregrines tended to take birds from the bottom, edge or tail end of the flock. Juvenile Dunlins were less proficient than the adults at drawing closely together to find safety in the middle of a flock, thus rendering them more vulnerable to Peregrine predation.

Overall, however, elements of the so-called life/dinner principle may come into play: it is more important for the prey to have evolved an ability to escape the predator than it is for the predator to succeed (invariably) in capturing the prey; of course ultimately the predator will die if it cannot kill and eat but it may be able to afford to fail in its hunting attempts, say, nine times and yet survive by catching prey on the tenth attempt. However, the prey species cannot afford to fail to escape, not even once.

5 | Breeding cycle

The Peregrine's annual breeding cycle across the planet spans varying months of the year, depending on the latitude at which the birds are to be found and whether they are in the Northern Hemisphere or the Southern Hemisphere. One could say that the Peregrine has two separate seasons, a non-breeding autumn and winter season and a breeding spring and summer one. Let us call them the 'off-season' and the 'on-season.'

In the British Isles, for example, broadly speaking the Peregrine's off-season begins around September when the young birds of that year become independent of their parents. It continues through the four months October–January, during which there may be some migration (or at least short distance movements) from the breeding grounds. Most adult Peregrines in Britain and Ireland remain in their breeding areas during the off-season (those in prey-poor habitats are more likely to leave) but they may move away temporarily when bouts of severe winter weather strike.

The on-season starts in February or March, when breeding Peregrines are seen more regularly at their nesting haunts, engaging in courtship and territorial activity. Eggs are laid in April (occasionally in late March), while May and June are the months when young are in the nest. These young birds are on the wing in July but are dependent on and fed by their parents for the next two months through to September. Breeding location, its altitude, local food supply and prevailing weather conditions may have an influence (in terms of days, occasionally weeks but not months) on those timings.

Previous page: A female Peregrine returns to its eyrie, in this case a disused Northern Raven nest. The sticks on the outer edge of the nest are visible.

Opposite: In years of good breeding conditions Peregrines may rear broods of three young. These recently fledged youngsters eagerly await the return of a parent with prey.

Territories and ranges

It is worth going over a few basic terms that are used to describe aspects of the Peregrine's breeding locations. The word 'territory' is often used for the whole area frequented by a breeding pair of Peregrines (or other raptors) but 'home range' is a better description for this. A true territory is the area a bird defends against others of its kind, but with the Peregrine only part of a home range is actively guarded in this way.

It is difficult to define the boundaries of a home range and hence its extent, but a territory should be easier to map out as it covers a smaller

area, and within much of it there is a good chance of witnessing actual territorial activity. 'Nest site' is not to be confused with territory or home range. It constitutes the nest and its immediate surroundings, in the case of the Peregrine typically on a cliff ledge. The term 'eyrie' (in earlier times 'aerie', derived from medieval Latin meaning an open space) is often used, especially in the case of cliff-nesting raptors such as the Peregrine, for the actual nest structure or part of a nest ledge in use in any one year. Eyrie has something of a poetic ring to it and I prefer it to the more prosaic 'nest.'

A Peregrine home range almost invariably contains a number of separate nest sites on one or (usually) more separate crags at varying distances from each other. Very occasionally, if there is just one small crag in a home range there may be only one nest site (a suitable ledge for breeding) on it. In one such instance known to me the nest ledge (about one metre in length) nevertheless holds two separate eyries as there is a scrape (of which more later) at each end of this ledge. In one small lowland Scottish glen there are ten alternative nest sites, spread across seven quite diminutive crags.

The less often used term 'nesting range' implies the area containing all of the alternative nest sites thought to be located within the home range of a single Peregrine pair. A home range is treated as such on the basis of data from a sequence of breeding seasons, showing that only one nest site or nesting range in a given locality was occupied by a breeding pair of the species in question in any one year.

Before we look at the annual breeding cycle in detail, let us examine the way in which pairs of nesting Peregrines apportion the habitat between them. For Peregrine census purposes, the counting unit is the territorial pair, but how does each pair fit into the overall structure of the breeding population, and what dictates the number of breeding pairs in any one region? Keys here are landscape (in terms of the availability of natural or man-made nest sites) and quality of the Peregrine's food supply. These two factors interact to limit the carrying capacity of the environment for breeding Peregrines (and may influence the numbers of non-breeding, non-territorial birds) and hence the number of breeding pairs that can exist in any one area. Take two regions of similar food supply for Peregrines, one with an abundance of suitable nest sites, but the other with far fewer. There may be no more breeding Peregrines in region 1 than in region 2, despite the additional nest sites in the former, since the territorial behaviour of its Peregrines restricts numbers to a lower level than nest site availability alone would indicate.

Peregrine breeding density varies widely in different parts of the world and within individual countries, depending on topography and food supply. To give an example, in the relatively productive (i.e. prey-rich) hill country on the southern fringe of the Scottish Highlands, typical breeding density averages around one pair per 5,250 hectares (13,000 acres) wherever there is a good supply of nesting cliffs. A different situation was discovered in central Scotland where in 1987 no fewer than five Peregrine pairs nested along 9.65km (6 miles) of an escarpment, giving a mean minimum distance of 1.93km (1.2 miles) between pairs. There, however, the Peregrines were able to range far and wide from the escarpment without the drawback of other nearby pairs having a stake in the available food supply.

Territorial behaviour

There are differing degrees of territorial behaviour among Peregrines, as Cade discovered in Alaska. There, a minimum radius from the eyrie, sometimes as little as 100 metres, was defended at all times by nearly all Peregrines. Beyond this radius there was a diminishing gradation of territorial aggression at distances of up to 1.6km (1 mile) from the eyrie. There can be still less aggression (perhaps none) against other Peregrines towards the outer limits of a pair's home range, which for hunting purposes may overlap with the ranges of other Peregrines.

It seems to be the birds' own behaviour towards each other that fixes, in some subtle way, the number of Peregrine pairs that can breed in a given area, assuming the essentials of availability of nest sites and an adequate food supply. If prey populations increase, the numbers of breeding Peregrines can rise through increased tolerance of closer nesting distances between pairs. Likewise, if food supplies decrease below a certain level (this level is difficult or impossible to define precisely), increased intolerance between pairs can lead to a decline in breeding numbers. Such intolerance is generally not obvious to the human observer (the birds must have a language, so to speak, among themselves on this) but it appears to form the basis of Peregrine relationships within the population of any given area.

Sometimes, however, Peregrines' behaviour towards each other is far from subtle. Actual violence may ensue, particularly in the case of migrant populations returning to their breeding quarters. Male-on-male and female-on-female fights occur readily when late-arriving potential breeders encounter opponents which have been on territory for two or three weeks. A Canadian study of urban Peregrines at Edmonton, Alberta, provided excellent opportunities for observing such behaviour. Ground fighting between the birds (with its attendant difficulties of disengaging from combat) was found to be especially dangerous for one or both adversaries, to the point of severe injury or even death. It was thought once that such behaviour is rare but it has now been shown to occur regularly, at least in this Peregrine population.

The breeding pair

Of course, relationships within the unit of the Peregrine pair are different, an essential first step being mutual acceptance between the two birds. This is hardly surprising (we act likewise in our own relationships), but in order to achieve this happy result Peregrines go through some distinctive rituals. Throughout the off-season, both members of a pair may remain together, for much of the time anyway. Peregrines on wintering grounds that are separate from their breeding haunts appear to be generally solitary, although pairs are seen at times. If a pair remains in its (breeding) home range over the winter, the male and female can appear to be singletons if frequenting separate cliffs for roosting.

Peregrine pairs that are established by the end of the off-season may

Opposite: Copulation, with its pair-bonding as well as reproductive function, can take place both before and after egg-laying. When seen mating, the size difference between the larger female and the smaller male Peregrine is obvious.

comprise the same two individuals that consorted in a previous year, or birds that had been strangers to each other, either a resident pairing with a former non-resident, or two non-residents joining up to fill a vacancy in the breeding population.

Courtship and pair-bonding

Although some autumn and winter pair-bonding has been noted, the process gets underway properly around mid- to late February. Jockeying for position at the nesting crags within a Peregrine territory (the eventual crag for the season may not have been chosen yet) occurs. Various potential mates are seen and accepted or discarded by either sex before the pair is formed, and is ready to face the rigours of the forthcoming breeding season. Pair formation is not so much just through male meeting female, but also requires the essential ingredient of a good breeding cliff, to which individual Peregrines are inexorably drawn as if by a magnet.

Until recently nest crags could be divided conveniently into first-, second- and third-class cliffs, basically in terms of height and hence security of the eyrie. More recently this division into categories of (natural) cliff has become somewhat outdated, with many more small crags and occasionally even ground sites now being used successfully for nesting.

Following pair formation, the Peregrine pair's distinctive rituals early in the on-season include mutual roosting on the nest crag, co-operative hunting by the pair, courtship flights, courtship feeding, what have been described as 'familiarities' on the nest crag, copulation and nest scraping. These activities may follow each other in (approximately) this order but to some extent also they operate simultaneously. Mutual roosting involves the Peregrine pair perching on the same cliff, and increasingly in each others' proximity. Co-operative hunting sees the two progressing from chasing prey separately within the home range to joining together to pursue the same bird. Courtship flights produce spectacular display flying (especially by the male), which is intended for the other member of the pair but probably also sends a territorial message to other Peregrines, along the lines of: "This territory is already occupied, go elsewhere." Co-operative hunting by the pair gives way to the male's courtship feeding of the female, who by then is moving towards her more sedentary role of 'nest mother and nest guardian.'

For the next two or three months, the male will provide for the female most of the food that she needs. Familiarities on the nesting crag involve ritualised displays (accompanied by calling between the pair) that are geared to the selection of possible nest ledges. These displays have acquired various names describing their actions, for example Head-low Display by either sex, Male Ledge Display, Female Ledge Display and Mutual Ledge Display. In copulation the male Peregrine mounts the female, flapping his wings and balancing continuously, while she leans forward with wings slightly open. Copulation continues up to and occasionally beyond egg laying, implying that it has a pair-bonding function as well as a physically reproductive one.

Nest scrapes and egg-laying

Instead of building a nest structure, the Peregrine uses its feet to scrape a shallow depression on the eyrie ledge (much used scrapes from previous years can be deeper) to contain the eggs. Scraping, by either sex but as time goes on increasingly by the female, is also thought to have a pair-bonding purpose, so it counts as one of the ledge displays.

Again taking the British Isles as an example, egg-laying (in the nest scrape of the year finally chosen by the female Peregrine) starts around the end of March or early in April. A typical clutch is three or four eggs, appearing over the course of a week or just under. Therefore incubation usually starts about a third to a half of the way through April, its duration varying between 28–33 days. The young generally hatch towards mid-May, although quite often hatching is around the end of the first week of May and sometimes nearer the final third of that month. Peregrines which lose their eggs (naturally or through human interference) during the first 10 days or so of incubation often lay a

Most incubation is by the female Peregrine (as here) but the male takes his turn for shorter periods.

second clutch, usually about three weeks after the disappearance of the first one.

Incubation is by both sexes, with the male taking his turn during about 30 per cent of the daylight hours but not at all at night. Nest changeovers (when one bird takes over incubation and allows the other to go 'off-duty') can be interesting spectacles. A typical sequence is this: the male Peregrine flies to the nest crag carrying prey for the female; she spots him and leaves the eggs; the male hands over the prey in mid-air to the female (the so-called 'food pass') who departs, perhaps to somewhere a few hundred metres distant for her meal; the male slips on to the eggs to incubate; after perhaps an hour the female returns, landing on the eyrie ledge; a few seconds later the male stands up (sometimes after verbal abuse from the larger female who, eager to resume incubation, tells him to leave) and flies away, not to be seen again for several hours; and the female steps carefully over the eggs and settles down in an incubating position.

The female is almost entirely responsible for brooding small young, an essential activity at times of cold, rain (or snow) and strong sunshine. The male Peregrine's lesser size makes brooding of even very small chicks a difficult task.

Overleaf: Peregrine chicks, one-third grown at about two weeks old, are fed by their mother.

Rearing the chicks

Hatching of the clutch takes place over 48 hours, usually not more than this. The young are brooded to protect them against cold, rain (even at times snow) and strong sunshine, mainly by the female but to a much lesser extent, and for just a few days, by the male; he has difficulty in covering several young, even when they are very small. Brooding decreases in frequency as the days go by and usually stops when the young are aged three weeks or more. The young are clothed in white down until about three weeks old, when wing and tail feathers start to appear.

By the age of four weeks, a chick's brown juvenile feathering is becoming obvious, and at five weeks old the young bird is well feathered with most of the down gone. A week later, the juvenile Peregrine is ready to fly. It is active, wing-flapping vigorously and clambering around on and just off the eyrie ledge. By this stage the sight of a parent arriving with food in its talons produces a noisy response from the young. The male Peregrine had increased his prey capture rate as soon as the young hatched, the female alone feeding them at first but the male taking a small part in this duty thereafter. Once the young are four or five weeks old they are able to tear up prey brought on to the eyrie by one or other parent bird and to feed themselves.

At four weeks of age (some two-thirds of the way through the fledging process) the juvenile Peregrine's brown feathering is starting to appear.

Fledging

The juvenile Peregrine makes only short flights at first. Generally males leave the eyrie ahead of females from the same brood; the males develop more quickly than their sisters. After only a few days on the wing, the young bird is competent enough to receive an aerial transfer of prey, foot to foot from adult to young. By that stage also the young are starting flying practice, chasing each other in what looks like play, although presumably it has a serious purpose in achieving co-ordination of eye, wing, tail and foot. Such co-ordination is important since, as Baker put it: 'Everything he is has evolved to link the targeting eye to the striking talon.'

Food provisioning of young Peregrines once fledged can be a dramatic event. All is quiet at the nest crag, the adults are away and the young are waiting, perched on different sections of the cliff. Suddenly an adult appears in sight round the corner of the hill with the characteristic bulge visible beneath it, a prey item held in the talons. There is instant commotion as the young Peregrines launch themselves off the crag in great excitement, calling loudly and pursuing the adult to encourage it to hand over the prey. The adult obliges, one of the youngsters gets the prize

and the others continue chasing after the adult bird in the hope of more. That will come with the next prey delivery.

It is known that Peregrines seize and release pigeons for their young to catch, presumably as part of their training to be competent hunters. Other species too are used in this presumed training role. On one occasion a female Peregrine was seen carrying to its young (by then two weeks out of the eyrie) what turned out to be a newly fledged Carrion Crow. The adult Peregrine dumped its catch on a knoll frequented by the young, whereupon the crow, until then apparently lifeless, crawled off to hide in Bracken (*Pteridium aquilinum*) but flew away, seemingly unharmed, on my approach.

By the time that they are five weeks old young Peregrines are largely down-free and approaching the flying stage.

Provisioning the young

While much is known about which prey species are taken by breeding Peregrines, there is less information on the distances travelled by adult birds on their hunting forays. This is a worthwhile topic for further research. Various distances from the nesting crag for breeding season hunting and prey capture have been reported, varying between 13km

(8 miles) in the northern Scottish Highlands, 15km (9.25 miles) in central Europe, 18km (11.25 miles) in southern Scotland and 19km (11.75 miles) in Colorado, United States. At the other end of the scale, it was estimated that Peregrines in part of the central Scottish Highlands took about 70 per cent of their prey within 2km (1.25 miles) of the nesting crags, although females (more inclined to hunt after the young are about half grown and especially once they are safely out of the eyrie) made their prey captures up to 6km (3.75 miles) away.

The long-range hunting distances quoted (ascertained largely by tracking of radio-tagged Peregrines) may be extreme examples. Arguably, what are more relevant are the regular, shorter (but still quite lengthy) distances travelled by hunting Peregrines in the breeding season. How far a Peregrine with young to feed will need to fly will vary of course in accordance with prey numbers and availability in the home range. An instance that I witnessed in the 2010 breeding season could be quite typical, although this is speculation. A Peregrine, probably male, was seen carrying prey of about thrush size in a direct line towards an eyrie (out of sight and in an adjoining glen) containing at that stage four young about two weeks old. There was no other plausible destination for this bird. The prey capture would have been at least 6.5km (4 miles) from the nest crag in this instance. The Peregrine's hard work in carrying home heavy dead prey for the young is eased when it removes one or both wings and legs and the head of the captured bird.

Previous page: Urban Peregrine populations have been helped by the fact that the species has readily adopted man-made nest sites on tall buildings, whether they were designed specifically for the purpose or otherwise. This juvenile was photographed at a site in London.

Survival and independence

Many Peregrine families comprise two or three young at fledging (a long term average brood size is 2.5) although there are regular examples of single young and, less often, of broods of four. Instances of single young may reflect poor food supply or wet weather (directly impacting on the young but also impairing the adults' hunting abilities), or a combination of these factors.

Weather conditions, particularly when the young are small, clearly play a big part in success or failure of a Peregrine's nesting attempt. The exposed situations in which many Peregrines nest may have a bearing on this. A successful nesting is followed by a full two months or so during which the young are still dependent on their parents, for food and to some extent for protection.

There has been debate as to whether the adults eventually drive their progeny away, or whether the youngsters leave of their own accord, at the close of the on-season. The more plausible view is that juvenile Peregrines gradually detach themselves from their parents once they are competent enough to make their own way in the world. Seasonal (autumnal) decreases of prey on the breeding grounds could play a part also in encouraging the young to leave, on the basis that the home range can still support an adult Peregrine pair but not both the adults and the young birds of the year.

Once the juvenile Peregrines are independent and on their way to new quarters, what are their immediate prospects, and for how long are they likely to live? There is a wide discrepancy between the life expectancy of established, street-wise adults and that of the inexperienced young birds. At the start of the off-season, the Peregrine population is almost at its seasonal high point, though a few young will have died already while still dependent on their parents. The year's young Peregrines have augmented the overall population (both territory holders and non-territorial 'wanderers') and have boosted the latter sector of the population in particular.

Numbers are then progressively depleted from their post-fledging high point (with a large part of this mortality thought to fall on the less experienced juveniles) until the on-season comes round again and with it the scope to augment the Peregrine population once more. By then much of the annual juvenile mortality may have occurred already. This seems to be the general picture for stable Peregrine populations. It is notable how stable such populations have been over the years, notwithstanding the vicissitudes described in chapters 7 and 8. Within this overall picture, "the annual balance sheet in numbers" as Ratcliffe described it, one can pick out some pointers to Peregrine lifespan.

Lifespan

Much more is known about the survival rates of adult than juvenile Peregrines. Nevertheless, an analysis of British and Irish ringing (banding) recoveries from 1923 to 1975 indicated a mortality rate of 30 per cent for first-year birds but rates of 25 per cent and 19 per cent for second-year and adult Peregrines respectively. Some caution may be in order here as Peregrines were experiencing major fluctuations in their fortunes over this timescale.

A 1977 to 1982 Peregrine capture and marking study in south-west Scotland produced estimated annual mortality of 11 per cent among established breeders, implying that average life expectancy of this part of the Peregrine population was a further 10 years after the first year of breeding. It was suspected that this mortality level was low by comparison with that of Peregrine populations in some other, less favourable districts where pressures of one sort or another on the birds were probably greater.

Nevertheless, continuing research in southern Scotland and northern England, from 2002/2003 onwards and based on the use of PIT (passive induced transponder) microchips on leg rings, has estimated annual mortality (within the breeding segment of the population) of 25 per cent for adult male Peregrines and 22 per cent for adult females; most of these Peregrines remained on the same territories year after year. By contrast, some Peregrine territories in this study showed particularly high turnover rates, probably due to criminal persecution of the birds; in the worst such case a minimum of three different males and three different females were recorded in a 5 year period. Canadian studies of the *tundrius* and *pealei* subspecies showed a mortality rate of 25 per cent for the former and annual turnover of 32 per cent for the latter.

Findings to date from the southern Scotland and northern England research disclose that the mean maximum ages of known-aged males and females were 8.8 years and 10.4 years respectively, the oldest birds of each sex however being a 17-year-old male and a 14-year-old female. An interesting parallel example is the celebrated female Peregrine present at the Sun Life building in Montreal, Canada, from 1937 to 1952, which

was at least 18 years old when she disappeared. These seem to be exceptional ages for Peregrines, most having rather shorter lives.

Turning to age of first breeding, there are indications from North America that female Peregrines usually come into full breeding condition at two years of age (some not until a year later) but that most males are not in such condition until they are three years old and sometimes not until four or five years of age. It is thought that a male Peregrine may need an extra year or two of experience before he can take on the role of principal provider for his mate (during much of the breeding season) and for his offspring.

A breeding season aberration is the dispossession of nesting Common Kestrels by Peregrines. Four instances of this bizarre behaviour were recorded in south-west Scotland in the 1960s and 1970s, attributed to Peregrines in the organo-chlorine pesticide era (see chapter 8) breaking their thin-shelled eggs and, presumably still charged up with the necessary hormones, having the urge to take over the Common Kestrels' nesting duties. In two of these cases the kestrels are known to have fledged. A similar event occurred in the southern Scottish Highlands in 2011, when a Peregrine pair fostered a brood of 4-5 Common Kestrels and reared these to the flying stage. It is thought that one or more of the kestrels (probably abandoned by their foster parents once they had left the nest) survived for at least 10 days thereafter but only by teaming up with an adjoining Common Kestrel family.

A juvenile Peregrine, recently fledged, keenly surveys its surroundings.

6 | Neighbours, enemies and friends

Peregrine interaction with other species has generally been better studied in the breeding season than outside it. In recent years, however, more has become known about such interaction in the off-season and away from the Peregrine's nesting haunts. Clearly, the Peregrine has neighbours of one sort or another throughout the year, although these may be more transitory in the off-season than they are in the on-season and within the bird's home range.

The neighbours fall into four categories. There are those that seem to be disregarded by the Peregrine or are too small to be noticed by it anyway; the prey species (whether regularly or only occasionally taken by Peregrines); competitors (some of which may become prey also at times); and, in the case of a few species only, potential predators. In a sense, many more of the Peregrine's neighbours are important to it than is immediately apparent, since the Peregrine depends fundamentally, through the food chain, on lower forms of life (such as invertebrates) to nourish the prey species on which it feeds. In other words, these lower forms of life ultimately dictate the Peregrine's presence or absence and, if present, its reproductive success or failure; there can be no Peregrines (especially breeding ones) if the food supply is inadequate, regardless of how many suitable nesting crags there are.

Many of the prey species often seem to take no notice of the Peregrine's presence, but that could be a false impression, as these species might suddenly take avoidance or defensive action if they perceived a hunting attack on them to be under way or imminent. Here we hark back to Treleaven's distinction between low-intensity and high-intensity hunting, relevant if the prey can distinguish between the two.

The competitors may compete with the Peregrine for food, for living space (especially nest sites) or for both. The most obvious, although arguably less fundamental, neighbours are the larger predatory ones. The Peregrine has very few life-threatening natural enemies, but some of the larger raptors (and owls) and a few mammals, in Europe notably the Pine Marten (*Martes martes*), are among these. The phenomenon of intraguild predation, whereby smaller predators may be restricted in

Previous page: An immature Golden Eagle (foreground) and a Northern Raven. Both have strikingly similar distributions to the Peregrine. Both are competitors with the falcon and the eagle is an occasional predator upon it.

their numbers and distribution by larger ones of the same kind, applies in this situation.

Peregrines and eagles

The Golden Eagle is the pre-eminent competitor against (and occasional predator on) the Peregrine over its wide Eurasian and North American range. Both inhabit the same hill and mountain country, so their paths are likely to cross, but there is strong evidence that the falcon, as the smaller bird, avoids the eagle if it can. A radio-tagged female Peregrine in south-west Scotland on hunting flights seemed to steer clear of the general vicinity of a Golden Eagle nest 6km (3.75 miles) from her own eyrie. On one occasion, when returning from a distance of 15km (9.25 miles), she avoided flying directly over the eagle nest even although that would have been the shortest route back to the Peregrine nesting crag.

Clearly the longer way home taken by the Peregrine was worth the extra expenditure of energy, as Golden Eagles are known to rob Peregrines of their kills. A falcon/eagle dispute, especially when the Peregrine is burdened by carrying prey, is best avoided by the falcon. Moreover, the Golden Eagle is the winner in competition with the Peregrine for the best nesting crags. This was demonstrated in south-west Scotland, where after 1945 Golden Eagles returned (four separate pairs consecutively) to breed in their erstwhile haunts and excluded Peregrines which had been nesting on the same crags. Peregrines were able to come back to these crags when two of the Golden Eagle pairs dropped out, apparently as a result of land use change.

There may be competition between both species for food, some of the Peregrine's larger prey species being regular components of the Golden Eagle's diet also. There are a few records of Peregrines, probably caught

A Golden Eagle nest site in the Scottish Highlands. Studies show that breeding Peregrines keep their distance from established Golden Eagle pairs.

unawares, being killed by Golden Eagles. In one case in the Scottish Highlands, examination of an adult Peregrine found at a Golden Eagle feeding perch showed that it had been killed by the eagle and not taken as carrion.

At or near its eyrie, the Peregrine reacts vigorously to a Golden Eagle's appearance, presumably because it recognises the eagle as an undoubted danger to itself and to its young, whether these are on the eyrie or have flown from it. Mobbing of Golden Eagles by Peregrines can be particularly intensive, as on one occasion in the southern Scottish Highlands when an eagle flew in front of and close to the falcons' nesting crag. First the male Peregrine and then the female, coming off her eggs at the first hint of trouble, flew at the Golden Eagle which alighted on the ground near the crag. The eagle was mobbed again by both Peregrines, jumping up and down whenever either of them passed close overhead. It then flew away, pursued by the falcons for about 1km (0.6 miles) with the male Peregrine stooping at it continuously.

It may seem from encounters such as this that the Peregrine 'wins' against the Golden Eagle, but from an ecological perspective it is the eagle that is dominant, as shown for example by its exclusion of Peregrines from the best nesting crags.

Unsurprisingly, other large eagles are often the Peregrine's enemies. At the British Columbia estuary mentioned in chapter 4, harassment of Peregrines by Bald Eagles occurred regularly, and apparently affected the Peregrines' use of parts of the estuary for hunting. Incidents of successful piracy by the Bald Eagles were recorded, with the Peregrines forfeiting to the eagles large ducks (of unknown species) which they had caught on three occasions. Bald Eagles on that estuary frequently interfered also with Peregrines when they were hunting Dunlins.

Other birds of prey

Eagle threat to Peregrines may take the form of interference with hunting patterns more often than of actual killing, but with the large owls the reverse seems to be the case. In Europe the Eagle Owl (*Bubo bubo*) and in

North America the Great Horned Owl (*Bubo virginianus*) are known predators of Peregrines. These owls, of course, have the advantage of being able to operate more effectively at night than Peregrines can, and have the potential to reduce Peregrine populations below levels that they might otherwise attain.

Opinions will differ but one can take the view that this is just an example of nature at work, and that we should not interfere provided that the dominant species is present within part of its natural range and has not been introduced (as opposed to reintroduced) through human agency. Eagle Owl and Great Horned Owl predation operates especially against incubating or brooding Peregrines and their young.

As the largest of the world's falcons, the Gyr Falcon is physically capable of killing a Peregrine, but most interaction of the two species seems to be in relation to competition for nest sites and to some extent for food. Nevertheless, in a curious incident in arctic Russia an immature Gyr Falcon was known to have killed one of three recently fledged young Peregrines. The tables were then turned, as shortly thereafter the Gyr Falcon itself was found dead, apparently having been set upon by one or both of the parent Peregrines.

Turning to other large to medium-sized raptors, Scottish Peregrines at their nesting crags seem to pay little attention (at least in some regions) to the reintroduced Red Kite (*Milvus milvus*) which may be perceived by the falcons as basically a carrion feeder unlikely to pose a threat to them. Peregrines react more obviously to Common Buzzards, seeing them perhaps as offering some risks to their young. Common Buzzards have been recorded occasionally as Peregrine prey species, as mentioned in chapter 4. The female Peregrine is a bit heavier than the average female Common Buzzard and markedly heavier than the average male.

Moving across the Atlantic, in western North America various facets of Peregrine/Prairie Falcon interaction have been recorded, the Prairie Falcon being the slightly smaller and lighter bird of the two. While in many situations Peregrines and Prairie Falcons live side by side without obvious hostility between them, it is usually the Peregrine that wins any

competition that there may be for nest sites. Peregrines are known to have killed Prairie Falcons on a few occasions but the reverse appears not to have been the case.

The Peregrine can have an impact on smaller raptors other than through direct mortality, by making a location a much less congenial hunting ground for smaller species than it was before. At an estuary in Washington in the north-west United States, hunting Merlins were observed seven times more frequently in the period 1980–1988, when no Peregrines were seen at the estuary, than in 1999–2005 when (following recovery of Peregrine numbers) slightly more Peregrines than Merlins were recorded there. It is thought that the Merlins, vulnerable to predation by the larger falcon when out on an open estuary, avoided hunting in the presence of Peregrines.

Other neighbours

The Northern Raven is a special case in its interaction with the Peregrine. Both species are often found at the same nesting crag or series of crags, so the student of one species can easily follow the fortunes of the other. Indeed, the Scottish Raptor Study Groups have named the Northern Raven the 'honorary raptor' for its close ecological relationship to the Peregrine. These Groups record the Northern Raven's numbers and breeding success or failure accordingly.

Northern Ravens show the same regularity of spacing between nesting pairs as the Peregrine. While the two species may nest some hundreds of metres apart (and generally seem to prefer to do so where there are a number of suitable nesting crags available) there are instances of very close nesting, sometimes within only 10–15 metres of each other. The Peregrine and Northern Raven indulge in frequent, noisy skirmishes (the latter has a distinctive and harsh anti-Peregrine alarm call) and seem to be bitter enemies, making the best of a bad job as it were of having to live and breed close together. There are recorded instances of each species killing the other, usually it seems in aggressive territorial-type encounters, but there have been cases of Northern Ravens becoming Peregrine prey, as stated in chapter 4.

The Golden Eagle is one of the few bird species which is dominant over the Peregrine.

Both species need the same sort of nesting crags and are drawn together to the best ones in the vicinity, to the extent that these are not already tenanted by Golden Eagles. Thus Peregrines and Northern Ravens need the same living space, but perhaps also they need each other as friends. Research carried out in the Trento region of the Italian Alps points in this direction. There the two species tended to use the same nesting crags, notwithstanding an abundance of suitable cliffs in the area. It was thought that the Peregrines deliberately associated with the Northern Ravens in order to get advance warning from them of potential avian and terrestrial predators. There seems no reason why that situation should not work in reverse, with the Northern Raven being warned by the Peregrine when danger appears to threaten.

Other apparent friends of the Peregrine in various parts of the world are some ground-nesting birds. One such situation was seen in Alaska where there was an evident association of Canada Geese (*Branta canadensis*) and Peregrines nesting in close proximity, the geese giving warning of approaching predators and the falcons assisting by driving them off.

7 | Peregrines and humans

One can only guess as to when the relationship between falcons and early humans, including both our own species (*Homo sapiens*) and its predecessors, first took shape. Since some other species engage in piracy there is no reason to suppose that humans did not do likewise, watching hunting falcons and seizing the opportunity to rob them of their prey on the ground before it could be consumed or carried away. During how many generations of our own kind did we take advantage of wild falcons' kills, before learning to catch the birds themselves and to train them to hunt for us?

Previous page: The Peregrine is the archetypal falconer's bird for flights after quarry in open country.

It has been suggested that hunter-gatherer peoples probably did not have the resources that would have been required to keep falcons in captivity, but that pastoral nomads were well placed to do so. The origins of falconry, initially perhaps only for securing food but subsequently for sporting purposes also, were very likely in the Far East (China and Mongolia) and in the Middle East, and date back to perhaps around 4,000 years before the present. It seems that from then on humans had a particular interest in falcons, although the first archaeological evidence of falconry (also from the Middle East) is thought to stem from around 2,750 years before present.

Falconry flourished from about 1,500 to around 400 years before the present, particularly in the feudal societies of European Christendom and Islam. In at least the British tradition of the time, the Peregrine was reserved for princes, dukes, earls and barons, or at any rate was seen as their favoured species. Golden Eagles and Gyr Falcons scored rather better, being allocated customarily to emperors and kings respectively. In practice, it is very likely the Peregrine was the more useful falconer's bird, even if it did not carry the status of the larger raptors.

Obtaining Peregrines for falconry

In some cases, the incentive to protect the Peregrine was linked to the system of estate rentals in operation at that time, which might include a requirement on the landholder to deliver a given number of young birds at regular intervals. There could be severe penalties, physical and financial, on anyone who stepped outside the rules protecting Peregrines. The law might be invoked, through the courts system of the time, to safeguard Peregrines in the sense of limiting the taking of their young.

In one instance the culprits seem to have got off lightly, perhaps due to their status in society. The maintenance of Dumbarton Castle (now a ruin) on the River Clyde was supported by revenues from, among other sources, an island off the west coast of Scotland which held breeding Peregrines. In 1609 the Constable of Dumbarton Castle raised a court action against two local dignitaries (including one Robert Hunter) for unlawfully taking Peregrines from the island. The Privy Council ruled: 'All the hawks quhilk bred on the said isle do properly belong to the king, and ocht to be forth command to his majestie....and discharges the said Robert Hunter and all others from meddling therewith.' Quite rightly, we now find removal of young Peregrines from their eyries as reprehensible. On the other hand, in the heyday of falconry the species enjoyed a privileged and preserved status, albeit with what was probably a limited take of young birds. That was not to last.

Clearly there was at that time a well-developed system of falconry practice, once Peregrines had found their way into captivity, allied with the controls that applied in relation to their taking. Nowadays, with the advent of captive breeding, the means of providing birds for falconry have altered radically. In former times Peregrines had to be secured by the straightforward if dangerous practice of climbing down to their eyries or, in the case of wild-flying birds, by more elaborate methods.

One such method used in The Netherlands (its techniques would be illegal now) was described by Gerald Lascelles 120 years ago, and is worth repeating for its historic interest. It was centred on what was described as 'a vast wild plain or heath' at Valkenswaard (note the place name) near Eindhoven, and involved live trapping of 'passage hawks' i.e. juvenile Peregrines on migration, for falconry purposes. By tradition, extending over many generations, several local families acted as Peregrine catchers, their resulting booty being marketed post-migration at annual local 'falcon fairs.'

An adult Peregrine which has been illegally shot is shown by wildlife crime officer Sergeant Chris Hine in Yorkshire, England.

The techniques used by the catchers involved live pigeons, as baits, attached to strings which could be pulled to and fro to attract a Peregrine; a captive, hooded Peregrine with a bunch of feathers near to it (simulating prey) to provide further allurement; sentinels (usually two) in the form of tethered shrikes, thought to be Great Grey Shrikes (*Lanius excubitor*) or Lesser Grey Shrikes (*Lanius minor*) but described only as 'grey shrikes', to give advance warning (through an exhibition of excitement and dismay) of a wild Peregrine's approach; and a system of nets (controlled by the catcher) for entrapping any Peregrine enticed down by the pigeons. In the interests of animal welfare, the shrikes had 'a little house' in which to shelter. Another necessary feature was a primitive sunken turf hut, where the falcon catcher remained hidden "with a good stock of tobacco and some occupation such as net-making or cobbling to while away the many weary hours of waiting" until such time as the shrikes alerted him to the imminent prospect of catching a Peregrine.

The era of persecution

Thus the Peregrine especially (but other raptors also) enjoyed a venerated status for centuries, far removed from the treatment accorded to them thereafter, especially in the so-called civilised world. The main impetus for that treatment was the fact that quarry species could be obtained more easily by the use of 'vile saltpetre', the main constituent of gunpowder, than by the more labour-intensive keeping and flying of captive raptors, and by the development of the shotgun, in particular the breech-loading variety. As late as 1551 in Scotland, killing of game birds with guns was illegal (in the interests of encouraging the sport of falconry and hawking) but before long the valued Peregrine became the detested target. Ratcliffe described it thus: "The fowling piece nevertheless gave its owner a much simpler and more convenient method of taking game and with it the realisation that all birds of prey were now competitors for the game. The early shot guns were doubtless soon turned against the former feathered accomplices, and the feud against the raptors began."

In Britain, a principal complaint against the Peregrine was (and in some quarters still is) its predation on Red Grouse, the main game bird of the uplands. Deforestation provided good opportunities for the Peregrine, and no more so than on the moorlands of north and west Britain. There the burgeoning semi-natural habitat of heather moorlands (Ling Heather was part of the understorey of the former woodlands) was able to sustain an abundant supply of Red Grouse, a staple prey species for Peregrines. The problem, at least from the wildlife conservationist's perspective, was that grouse moors were not established in order to benefit Peregrines. Records show that Peregrine destruction was intense over much of the country (in coastal and lowland areas as well as in the hills) for some 150 years, in the 19th and first half of the 20th centuries.

It is a mistake to assume that all those with an interest in game bird shooting were inveterate destroyers of Peregrines. At the height of the killing time A. E. Gathorne-Hardy described late 19th century shooting forays in Argyleshire in western Scotland and wrote of the Peregrine: "I never fire my gun at the magnificent birds, and rejoice at the laird's orders that they should not be trapped. Inveterate poachers they are, no doubt – but what a wonderful thing is the swoop of a wild peregrine!"

That was written from the landowner's and shooting tenant's perspective, but it was (and in some places still is) the employee, i.e. the gamekeeper, who did most of the actual killing. Here too, not all attitudes were identical. Colin Gibson's biography of a gamekeeper in the eastern Scottish Highlands in the early 20th century has him expressing serious doubts about including the Peregrine on his 'vermin list' and includes a quote by a neighbouring gamekeeper that "the peregrine was the natural enemy of game birds, but it would never empty a moor or spoil the stock, any more than an otter would spoil the stock of a salmon river". The subject of the biography may have been encouraged partly through witnessing Peregrines killing Carrion Crows or Hooded Crows (*Corvus cornix*), arch enemies of the gamekeeper at nesting time.

Nevertheless, when game preservation was at its most intense, tolerance of the Peregrine seems to have been a rare commodity. It was not too difficult to shoot or trap the birds at their nest crags, nor to trap or poison them at kills if these happened to be found. Leaving aside verbal and written evidence, the signs on the ground have been all too obvious, in the shape of disused trapping sites (flat stones, sometimes with the wires for securing the traps attached and occasionally even the traps themselves) on top of nesting crags, and the remains of stone hides, below the crags and within 12 bore shotgun range of nest sites.

To be fair, many (not all) such relics seem to have been left behind by previous generations of gamekeepers and not by the current generation. Increased interest in Peregrine monitoring saw many of these structures removed or rendered inoperative. Some remain, in one case on a favourite Peregrine perching place an unset and therefore harmless steel gin trap attached to a small stake. With the passage of time the trap and stake have been totally grassed over and will probably never be seen again, but they remain there as a curious although typical example of gamekeeping archaeology. Legend has it that a previous gamekeeper (pre-1939) trapped the male Peregrine at this location every year without fail, replacement males being drawn inexorably year after year to this regularly occupied territory.

More subtle evidence of illegal interference with Peregrines is provided by studies comparing the bird's presence or absence, and (if present) its breeding success or failure between different landholdings and varieties of land use. A study examining breeding success of Peregrines in three separate Scottish regions between 1991 and 2000 disclosed that nesting success was significantly lower on land managed primarily for Red Grouse shooting, as opposed to land used mainly for forestry or hill sheepwalk purposes. Yet the former land use category provided Peregrine habitat that was at least as good as the two latter categories.

Writers on Peregrines

Fortunately, amidst these problems the Peregrine has always had its literary admirers, although probably they had little effect on the prevailing anti-Peregrine (and general anti-predator) sentiments of the persecution era at its height. Before the game preservers' main onslaught on the Peregrine began, Sir Walter Scott brought the bird into his famous *The Lady of the Lake*, published in 1810. This epic poem starts with an energetic and noisy royal stag hunt along the southern fringe of the Scottish Highlands, one of Scott's favourite haunts and still good Peregrine country today. Scott wrote: "The falcon, from her cairn on high, Cast on the rout a wondering eye, Till far beyond her piercing ken, The hurricane had swept the glen." Did Scott merely have a vivid imagination, or did he have detailed knowledge, in advance of his time, about the whereabouts of Peregrines and other species? In this instance it is not difficult to work out which Peregrine crag Scott had in mind if he was indeed 'in the know', but there may have been much historical knowledge, potentially useful to the modern Peregrine recorder, which died with its erstwhile custodians.

Some writers who concentrated on other species have brought Peregrines into their stories, clearly captivated by their powers of flight. Henry Williamson wrote about his adopted county of Devon in south-west England in the immediate post 1914–1918 war period, a time when the Peregrine was still Public Enemy Number One (probably as much in Devon as elsewhere), although it was given some respite in wartime when minds and guns were concentrated elsewhere. Williamson's best known work, *Tarka the Otter*, is on the face of it a single species book but vivid references to Peregrines are liberally scattered throughout, as where he describes the passing of a snow blizzard: "And the sky was to the stars again – by day six black stars and one greater whitish star, hanging aloft the Burrows, flickering at their pitches; six peregrines and one Greenland falcon. A dark speck falling, the *whish* of the grand stoop heard from half a mile away; red drops on a drift of snow. By night the great stars flickered as with falcon wings, the watchful and glittering hosts of creation."

Clearly there are gamekeepers who have been admirers as well as enemies of the Peregrine, witness the comment by one (as quoted by Ratcliffe) that "the Falcon is a real sportsman; I just grudge him the maintenance of his family".

Other conflicts

Some egg collectors and pigeon fanciers, too, have come into the category of admirers of the Peregrine as well as being its enemies. The 'eggers' have an obvious interest in wild birds (and some have been first-rate field naturalists), but when it comes to the taking of eggs theirs is a warped enthusiasm. Theft of Peregrine eggs was very much in fashion in the past. The species' comparative rarity gave added zest to the collecting mania, to the extent that in certain regions of Britain much, sometimes virtually all, of the potential Peregrine breeding productivity was nullified year after year through the activities of egg collectors.

At one time there may have been an element of justification for the collecting of eggs, to further scientific knowledge of one sort or another. Some eminent naturalists were collectors, or at least started their interest in the natural world in that way. Nowadays, egg collecting is seen, quite rightly, as a bizarre and damaging illegal pastime with no valid scientific purpose behind it, even if its proponents claim otherwise. The activity

Opposite: Pigeon fanciers were suspected of being responsible for laying spring traps at this Peregrine nest in the English West Midlands in 2008. An investigator from the RSPB is abseiling down to gather evidence.

Opposite below: Illegally placed spring traps have been found at several Peregrine nests in the West Midlands in recent years. This male eventually died from its injuries, while its mate disappeared. In such cases, the young Peregrines can sometimes be 'fostered' by another pair if they are discovered in time (see pages 110-111).

Above: Following the discovery of two orphaned Peregrine chicks at a nest site in the English West Midlands, RSPB climbers moved in to rescue them.

Opposite above: The chicks were then placed in two different foster nests, where they successfully fledged. The relocated chick is the slightly smaller bird on the right of the photo.

does not appear to be having a serious effect on Peregrine populations today, in many countries at least, but it is as well to be vigilant in case it should enjoy a resurgence. On a positive note, the egg collectors (or at least a few of them) did the Peregrine an unexpected service. That story is for chapter 8.

It is clear from conversations with individual pigeon fanciers that many have a keen interest in wild birds as well as in their own pigeons. Unfortunately, however, pigeon fanciers as a body cannot be described as anything other than hostile to the Peregrine. The racing pigeon is a domestic form of the wild Rock Dove (*Columba livia*) and has been bred so as to enhance its homing abilities, for return as quickly as possible to its home loft. Pigeon-racing involves transporting batches of the birds to release points (at varying distances from the lofts) and then timing them back to their lofts. Its popularity increased with the coming of the railways from around 1840 onwards which allowed rapid transport of pigeons. Now most movement to the release points is by road not rail.

Domestic pigeons (either racers or the many feral, i.e. gone wild, birds) form an important food supply for Peregrines. Inevitably, many racing pigeons are at risk of predation by the falcon when passing through or close to Peregrine country. It has proved difficult to estimate accurately the numbers of pigeons taken by Peregrines, although it seems to be low as a proportion of the racing pigeon population as a whole. Whatever the true figure may be, for many years pigeon racers have argued for relaxation of the law protecting Peregrines and certain other raptors. Some Peregrine eyries have been targeted directly and illegally. It is unfortunate that pigeon fanciers seem unable to accept predation by Peregrines and other raptors as something that should be tolerated as one

of several natural hazards to which all pigeons are exposed equally. A parallel is the racing yachtsman who has to accept the vagaries of the weather, a natural hazard and one that will potentially affect all other yacht racers equally.

While there have been and still are those factions and those individuals who can only be described as 'anti-Peregrine', fortunately there are now many more who are keen protagonists of the birds. This has been especially so over recent decades, when the Peregrine has been prominent in the public eye following its subjection to and recovery from the effects of organo-chlorine pesticides. The Peregrine's increasing tendency to nest on man-made structures provides good opportunities for viewing it at quite close quarters, leading to increased interest in the bird and its conservation. Moreover the species has had much positive media coverage, not least through some excellent television documentaries. The effort put into research and field study of the Peregrine has greatly increased our knowledge of its behaviour, distribution and population dynamics.

8 Future threats and opportunities

The saga of the organo-chlorine pesticides and the Peregrine is well known. As the outstanding threat to the species, the organo-chlorine story is a fundamental part of the Peregrine's history to date. It has been said that after the 1939–1945 war we left the Age of Metals (in which we had been living for 4,000 years) and entered the Age of Chemicals. The Peregrine was cast into that Age with a vengeance, with the arrival initially of DDT.

This chemical's properties as an insecticide were discovered in 1939, and the agricultural use of DDT commenced around 1947. The effect on the Peregrine from this persistent and thus (from the farmer's perspective) welcome substance was not recognised until more than 10 years later. It was mirrored by similar effects on some other species (especially bird-eating raptors) at the top of their food chains. The more toxic organo-chlorine chemicals aldrin, dieldrin, endrin and heptachlor (in the cyclodiene family of insecticides) came into use in Britain around 1956–1957, leading to increased mortality of Peregrines and various other birds and mammals.

The general picture in Britain appears to have been one of reduced Peregrine productivity (from eggshell thinning and egg breakage) as a consequence of DDT use, accentuated by rapid population decline brought on by the cyclodienes. Being fat-soluble, all of these chemicals accumulated at sub-lethal levels in the bodies of many prey species, from which the Peregrine ingested cumulative and ultimately damaging (to reproduction) or lethal doses.

Previous page: Studies of Peregrines, such as ringing the young, have made an enormous contribution to our understanding of the species. Of course, observers should never approach nest sites and should leave such work to a licensed ringer.

North America

It seems that in North America it was DDT use, leading to widespread reproductive failure, that was more directly responsible for Peregrine population decline. It is worth looking again at the pattern of Peregrine decrease (and, over wide areas, extirpation) and recovery in the United States. The Peregrine population in the eastern States at the beginning of the 20th century was put at about 400 pairs. By 1965 no nesting Peregrines were found east of the Mississippi River, and in 1975 only 39 breeding pairs were recorded in the western States out of the 300 plus pairs known there historically.

The picture thereafter is one of progressive recovery. Over 5,000 Peregrines were released in the continental United States (with a further 1,667 in Canada) between 1974 and 1999. DDT use, largely banned in Canada in 1969, was prohibited in the United States in 1972 where a ban on aldrin and dieldrin followed in 1974. Those steps, together with the reintroduction programme and augmented by a degree of natural recovery, led to more than 1,000 pairs being re-established by 1999. Almost all of the Peregrine recovery goals of individual States were met but in most cases were exceeded. The North American Peregrine recovery programme represents the largest and most successful attempt ever to restore an endangered species. It has provided the impetus and methodology for many similar raptor restoration programmes around the world and shows what can be done with the necessary level of determination, expertise and funding.

In Britain, an unintended consequence of the pigeon racers' lobbying for relaxation of the Peregrine's protected status (this came to a head in 1960) was the first national enquiry into the bird's population status, fieldwork for which took place in 1961 and 1962. If it had not been for that enquiry, the unprecedented, pesticide-induced decline in Peregrine numbers would have been discovered only later, perhaps too late for survival of the species in Britain. There the Peregrine decrease levelled out in 1963 and natural population recovery started in 1967, following successive but gradual phasing out of the most damaging chemicals, despite resistance from the agro-chemical and agriculture lobbies. This process of decline and recovery was echoed, with variations in time and scale and with the help of captive breeding and release programmes, in much of the developed world.

The egg connection

Seen from a British perspective, the service rendered to the Peregrine by egg collectors was the co-operation of some of their number in making

Peregrine eggs (centre) are frequently found within the collections of serious egg collectors.

eggs taken pre-1947 available for measurement for eggshell thickness, against thinner-shelled eggs acquired (for scientific purposes and under licence) at later dates. Concerns remain about organo-chlorine use in other parts of the world (as in the case of increased use of DDT in Africa, indoors but nevertheless possibly contaminating migratory prey species) but broadly speaking this problem seems to have been overcome in (especially) Europe and North America.

Even so, there is no room for complacency. We should be alert to the possibility of other chemical risks emerging (there are worries about, for example, environmental contamination caused by some industrial chemicals such as flame retardants), and we would do well to heed the lesson that the earlier and very serious chemical threat to the Peregrine was discovered largely by accident.

Deliberate persecution

It is essential to keep the lid on wildlife crime affecting the Peregrine. Recent case history in Britain is one of gains and losses in this respect. In the 1980s the level of egg and chick thefts of Peregrines was almost twice as high as for any other species, many of the nest robberies supplying an illegal falconry black market. A series of criminal convictions of Peregrine dealers in the 1990s (using DNA profiling to challenge false claims of captive breeding) plus improvement in captive breeding techniques by genuine falconers led to a significant decrease in eyrie robberies. However, a conviction in 2010 for attempted smuggling of Peregrine eggs out of Britain shows that the problem has not gone away.

Continuing routine criminal persecution of Peregrines by some grouse moor interests results in low densities of the bird or its absence from large areas of suitable habitat. A solution would be for game bird managers to move to a system whereby large numbers of game birds shot is no longer a prime objective. Interestingly, it has been argued that an answer to Peregrine predation on Red Grouse is another Peregrine, in that the species' territorial behaviour leads to a resident pair excluding intruder Peregrines from at least part of its home range. Other shooting, trapping and poisoning of Peregrines still occurs at or near some of their nesting sites, in locations where suspicion falls on the activities of pigeon fanciers. Pigeon fanciers need to accept Peregrine predation as an inevitable natural hazard.

In some countries, there seems to have been a resurgence of old problems such as the killing of Peregrines for taxidermy and subsequent sale of the specimens to collectors. In a wider wildlife management context, more attention should be paid to the now established general principle that preservation of top predators helps rather than hinders biodiversity conservation. Hopefully, public interest in the Peregrine will ensure that legislation protecting it will remain in force, notwithstanding the strident voices of a minority that would have it diluted.

Other hazards

Increased recreational use of the Peregrine's living space can lead to problems of inadvertent disturbance, at one level impacting on breeding success but at another, more serious, level affecting territory occupation and thus potentially reducing the overall breeding population. Liaison between wildlife conservationists and recreation participants (such as rock-climbers) can help in individual situations. The same problems (of deliberate interference with and inadvertent disturbance of Peregrines) manifest themselves, to varying degrees, in other European states apart from the United Kingdom.

There is a different, arguably more fundamental, negative influence looming on the horizon (in some places it is well above it), which is the adequacy or otherwise of the Peregrine's food supply. Hitherto in this book the Peregrine's recovery and conservation status have been shown in a good light, but in parts of the uplands of Britain, for example, there are warning signs. Successive national Peregrine surveys (at 10-year intervals) have shown that numbers there have been progressively declining, contrary to the trend elsewhere. Evidence points to a decrease in the Peregrine's prey species as the cause, at least for those regions where deliberate persecution of Peregrines has not been recorded or is little prevalent. Further research on this aspect of Peregrine ecology is needed, together with land use policies that enhance rather than deplete the species' food supply. The problem could manifest itself elsewhere, given the paramount importance of the Peregrine's prey base and the intense modern pressures on nature in today's world.

In general, vigilance is called for from all those concerned with the Peregrine's welfare, be they professional conservationists, amateur enthusiasts carrying out monitoring and survey, or interested members of the public who can bring pressure on politicians to take wildlife conservation seriously. It is the unanticipated threat that could turn out to be the most damaging. Ratcliffe put it well, referring to the pesticide saga, when he wrote: "I, for one, would not have the confidence to say that we can predict all the eventualities for threat. I prefer to see a lesson in the fact that the last and most serious threat to the Peregrine was largely unforeseen."

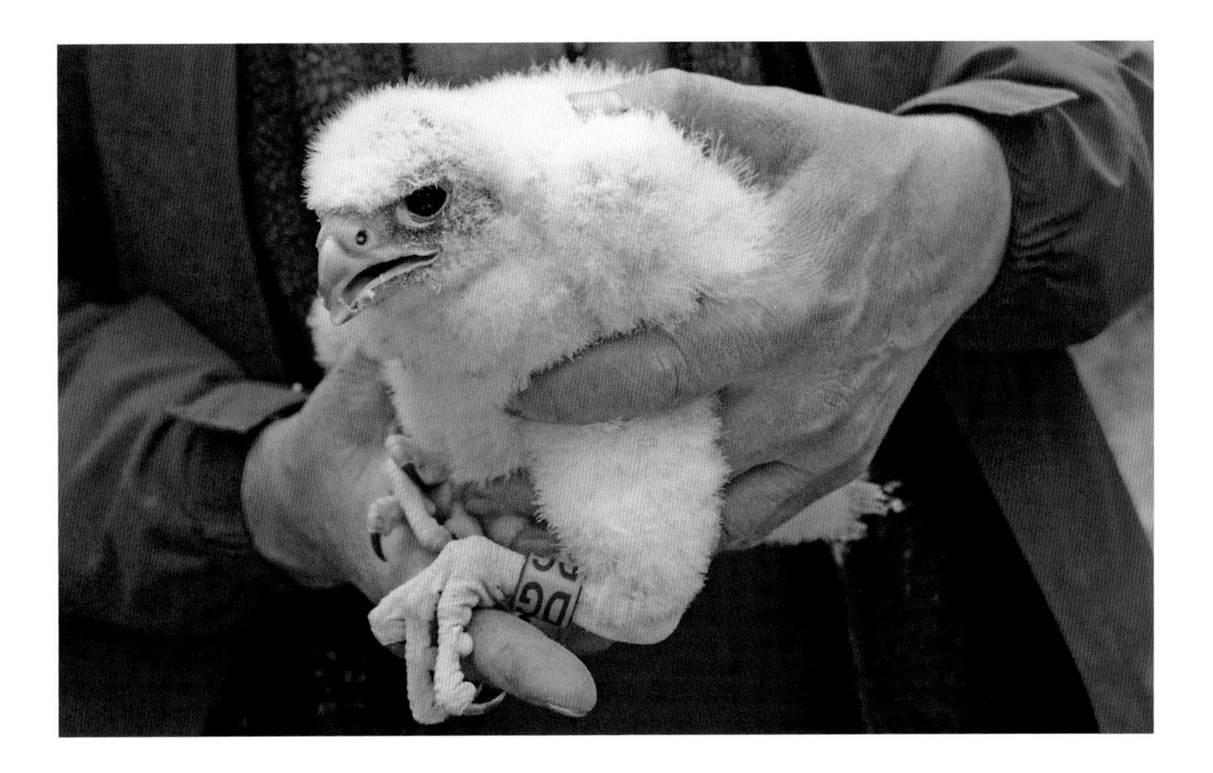

Research carried out on Peregrine populations, including results obtained from ringing, continues to be important.

Therefore, wariness as to the species' future is definitely needed. Yet at a time when so much of the natural world and its biodiversity is under siege, the Peregrine has proved to be both resilient and adaptable. The species survived the trauma of intense human persecution and the ravages of the pesticide era, damaged in numbers but with the potential for recovery. This potential has been realised to a considerable extent, partly as a result of the Peregrine's adaptability. Who would have predicted even two decades ago that by the 2011 breeding season there would be 24 Peregrine pairs within the M25 motorway encircling London, or that 21 pairs would be found in the predominantly lowland county of Shropshire – where only two pairs were recorded in 1991? Vigilance of the conservation community, the public's concern for the bird and an adequate level of environmental commitment by government should be capable of ensuring a bright future for the Peregrine.

9 Peregrine watching

RINE FALCON GLOSSARY
Korean:
Mongolian:
Thai:
Urdu:
Vietnamese: Cắt lớn
AFRICA
Afrikaans: Slegvalk
Swahili: Kozi tembere
Viking

With Peregrines again now widely distributed in suitable habitat in many countries, there are good opportunities to watch them at nesting haunts (both rural and, increasingly, urban) and, more randomly, to observe them in the countryside at large. Some of the best Peregrine watching now is at public viewing places, whether in town or in country.

A word of warning is needed: as the Peregrine is widely protected by law and is, for example, listed in the UK on Schedule 1 of the Wildlife and Countryside Act 1981, it is therefore an offence, and one that without an official scientific licence carries substantial penalties, to intentionally or recklessly disturb the bird while it is building a nest (in this case defined as, for instance, pre-laying nest ledge scraping) or is in or near a nest containing eggs or young, or to disturb Peregrine young that are still dependent on their parents. If you hear the '*cack-cack-cack*' alarm call, you are too close and need to back off. In any event, and by taking care to avoid disturbing breeding Peregrines, it is much more rewarding to witness the natural behaviour of birds that are not alarmed by close human presence, and that are going about their daily lives apparently unconcerned.

My Peregrine survey work has been largely restricted to inland, upland districts and this is reflected in the following description of Peregrine-watching in the breeding season. Others such as Treleaven have written vividly about sea-cliff Peregrines. His advice is useful, for instance that it is better to be a good distance from the eyrie and in comfort, not "fifty yards closer with a rock sticking in your back". Current advice would include avoidance of attracting sheep ticks to one's person. At first glance, the hill country of any one district may seem uniform, but to me each Peregrine breeding territory has a character of its own, with subtle differences in landform, vegetation and general ambience.

The few weeks from early March to about mid-April is a time for seeing Peregrines at their best, engaging in spectacular display flights at or around the crags, and going through the elaborate process of choosing the eyrie ledge to be used for the season. Peregrines can be quite vocal then, contact calling between them carrying some distance; listen for the '*klee-klutch*' call, perhaps the first indication of a hitherto unseen Peregrine's presence on a crag.

Previous page: Peregrine watching in cities is becoming more common as urban populations of the species increase. Here people at an event organised by the RSPB are looking at Peregrines on the Tate Modern in central London.

Today there are many opportunities to observe Peregrines, often through organized watches of nests which do not disturb the birds.

Thereafter life at the eyrie is quieter during the four weeks of incubation. The nesting cliff can seem deserted until the off-duty bird comes in to relieve the sitting one, often for about an hour only in the case of the male taking over from the female. That is when careful and consistent observation (in several layers of clothing to guard against the cold) may be needed, with eyes glued to binoculars sometimes for hours on end so as to pinpoint the current year's eyrie ledge. Nevertheless, a perched bird on the crag, particularly the whiter-breasted male, may be picked out from many hundreds of metres away, and can provide a clue as to whether or not a Peregrine breeding haunt is in use.

From mid- to late May, life at the nesting cliff livens up. The adult birds are more in evidence when they bring in food for the growing young during the six weeks from hatching to fledging. The final two weeks in the eyrie and the early part of the post-fledging period (before the young birds start to travel more widely within the home range with their parents) are rewarding times to watch Peregrines, to count large young standing on and moving around the eyrie ledge and to enjoy the sight and sound of these young in flight, calling and chasing after the adults for food.

At all stages of the breeding season (and indeed with Peregrine-watching generally) it is a great advantage to have the sun behind one. Since many inland Peregrine nesting crags face east (an accident of geology) it often pays to watch in the morning rather than the afternoon.

Resources

Books

Baker, J. A., *The Peregrine*, Collins, 1967.
Cade, T. J., *The Falcons of the World*, Collins, 1982.
Cade, T. J., J. H. Enderson, C. G. Thelander and C. M. White (eds), *Peregrine Falcon Populations. Their management and recovery*, The Peregrine Fund, 1988.
Cade, T. J. and W. Burnham (eds), *Return of the Peregrine*, The Peregrine Fund, 2003.
Crawford, J. H., *From Fox's Earth to Mountain Tarn*, John Lane, The Bodley Head, 1907.
Crawford, T. (ed.), *The Lady of the Lake*, The Association for Scottish Literary Studies, 2010.
Gibson, C., *Highland Deer Stalker*, William Culross & Son, 1958.
Gordon, S., *Highland Days*, Cassell & Company, 1963.
Hardey, J., H. Crick, C. Wernham, H. Riley, B. Etheridge and D. Thompson, Raptors: *A field guide for surveys and monitoring* (2nd edn.), The Stationery Office, 2009.
Hickey, J. J. (ed.), *Peregrine Falcon Populations. Their Biology and Decline*, The University of Wisconsin Press, 1969.
Lascelles, G., *The Art of Falconry*, Neville Spearman, 1971 (first published 1892).
Macintyre, D., *Nature Notes of a Highland Gamekeeper*, Seeley, Service & Co., 1960.
Michell, E. B., *The Art and Practice of Hawking*, The Holland Press, 1959 (first published 1900).
Newton, I., *Population Ecology of Raptors*, T & A D Poyser, 1979.
Ratcliffe, D., *The Peregrine Falcon* (2nd edn.), T & A D Poyser, 1993.
Sielicki, J. and T. Mizera (eds), *Peregrine Falcon Populations*, status and perspectives in the 21st century, Turul Publishing & Poznan University of Life Sciences Press, 2009.
Thompson, D. B. A., S. M. Redpath, A. H. Fielding, M. Marquiss and C. A. Galbraith (eds), *Birds of Prey in a Changing Environment*, The Stationery Office, 2003.
Treleaven, R. B., *The private life of the Peregrine Falcon*, Headline Publications, 1977.
Vesey-FitzGerald, B., *The Vanishing Wild Life of Britain*, MacGibbon & Kee, 1969.
Williamson, H., *Tarka the Otter*, The Bodley Head, 1965 (first published 1927).

Media and press

BBC Television, *Wildlife on One*, 'The Peregrine: nature's top gun', 2003.
The Field, 'Under the eye of a falcon', 30 January 1980.

Newsletter

The Hawk and Owl Trust, *Peregrine*, Issue number 93, Autumn 2010.

Organisations

The Raptor Research Foundation.
The Journal of Raptor Research.
The Raptor Research Foundation 2009 Annual Conference, Conference Programme Book.

Websites

BirdLife International: www.birdlife.org
BTO: www.bto.org
Global Raptor Information Network: www.globalraptors.org
The Peregrine Fund: www.peregrinefund.org
RSPB: www.rspb.org.uk

Scientific papers

Banks, A. N., H. Q. P. Crick, R. Coombes, S. Benn, D. A. Ratcliffe and E. M. Humphreys, '*The breeding status of Peregrine Falcons Falco peregrinus in the UK and Isle of Man in 2002*', *Bird Study* (2010) 57, 421–436.

Brown, J. W., P .J. V. C. de Groot, T. P. Birt, G. Seutin, P. T. Boag and V. L.Friesen, '*Appraisal of the consequences of the DDT-induced bottleneck on the level and geographic distribution of neutral genetic variation in Canadian peregrine falcons, Falco peregrinus*', *Molecular Ecology* (2007), 1–17.

Drewitt, E. J. A. and N. Dixon, '*Diet and prey selection of urban-dwelling Peregrine Falcons in south-west England*', *British Birds* (2008) 101, 58–67.

Mearns, R., '*The hunting ranges of two female peregrines towards the end of a breeding season*', *Raptor Research* (1985) 19(1), 20–26.

Mearns, R. and I. Newton, '*Turnover and dispersal in a Peregrine Falco peregrinus population*', *Ibis* (1984) 126, 347–355.

Mitchell, J., '*Peregrines and Man in Dunbartonshire*', *West Dunbartonshire Naturalist Report* (1984) 6, 5–10.

Mitchell, J. and R. A. Broad, '*Close nesting of Peregrines in Stirlingshire*', *Glasgow Naturalist* (1987) 21, 359.

Ratcliffe, D. A., '*The peregrine population of Great Britain in 1971*', *Bird Study* (1972) 19, 117–156.

Sergio, F., F. Rizzolli, L. Marchesi and P. Pedrini, '*The importance of interspecific interactions for breeding-site selection: peregrine falcons seek proximity to raven nests*', *Ecography* (2004) 27, 818–826.

Sergio, F., I. Newton, L. Marchesi and P. Pedrini, '*Ecologically justified charisma; preservation of top predators delivers biodiversity conservation*', *Journal of Applied Ecology* (2006) 43, 1049–1055.

Treleaven, R.B., '*High and low intensity hunting in raptors*', *Z. Tierpsychol.* (1980) 54, 339–345.

Treleaven, R. B., '*Note on the hunting behaviour of the Peregrine*', *The Falconer* (1982) 7(6), 398–400.

Photographic credits

Agami: Han Bouwmeester (pages 31, 54, 73), Marc Guyt (pages 10, 14), Edwin Kats (pages 8, 23, 52), Wil Leurs (pages 32, 36), Arie Ouwerkirk (pages 6, 38), Ran Schols (page 51), Reint Jakob Schut (page 30) and Kristin Wilmers (page 46); Eleonor Bentall/rspb-images.com (page 120); Dave Bromley/RSPB (page 109 below); John Chapman (page 4); Trevor Charlton/RSPB (page 104); Ed Drewitt (page 119); James Leonard/RSPB (pages 110, 111); Don MacCaskill (pages 43, 70, 84, 85, 95); Dave Morrison (page 86); Stig Frode Olsen (front cover, back cover, pages 2, 16, 18, 20, 24, 29, 56, 61, 62, 64, 66, 92, 99, 128); NHPA: Jordi Bas Casas (page 44), Robin Monchâtre/Biosphoto (page 13) and David Tipling (page 100); Photolibrary: Ralph Ginzburg (page 48); Terry Pickford (pages 27, 28, 77, 80, 81, 82, 112); Mike Read (pages 35, 91, 123); Shiva Shankar (page 40); Guy Shorrock/RSPB (page 109 above); Russell F Spencer (page 59); and Mark Thomas/RSPB (page 116).

Index

Right: Four images illustrating an immature Peregrine as it goes into a stoop.